AF845584

Getting Heard but Not Listened To

Democratic Dilemmas and Policy Responsiveness

Series Editors: Thomas Bryer and Egle Butkeviciene

Public policy issues divide societies around the world, with resulting tensions that threaten the fabric of social cohesion and certain norms of empathy, compassion, and civility. Today, such hot button issues as poverty, refugee integration, immigration, and others are the source for implicit and explicit debate concerning the "fitness" and "worthiness" of some individuals and groups of individuals to be one with society. In social and democratic terms, some citizens are included-in and some are included-out. This series consists of books that address the democratic dilemmas that arise with desire to be responsive through policy to competing values, interests, and positions. Special attention will be given to local governance, where divisions often manifest most clearly, but works that address higher levels of government, as well as supranational governments, will also be included in the series. Accepted titles will be sole- or co-authored monographs or edited volumes, with emphasis on deep case analysis, comparative case studies, comparative empirical studies, or theoretical explorations.

Recent Titles

Poor Participation: Fighting the Wars on Poverty and Impoverished Citizenship by Thomas Bryer and Sofia Prysmakova-Rivera

Getting Heard but Not Listened To: An Analysis of Public Participation in Environmental Impact Assessment (EIA) in Brazil By Cristiane Bená Dias

Getting Heard but Not Listened To

An Analysis of Public Participation in Environmental Impact Assessment (EIA) in Brazil

Cristiane Bená Dias

LEXINGTON BOOKS
Lanham • Boulder • New York • London

Published by Lexington Books
An imprint of The Rowman & Littlefield Publishing Group, Inc.
4501 Forbes Boulevard, Suite 200, Lanham, Maryland 20706
www.rowman.com

6 Tinworth Street, London SE11 5AL, United Kingdom

Copyright © 2021 by The Rowman & Littlefield Publishing Group, Inc.

All rights reserved. No part of this book may be reproduced in any form or by any electronic or mechanical means, including information storage and retrieval systems, without written permission from the publisher, except by a reviewer who may quote passages in a review.

British Library Cataloguing in Publication Information Available

Library of Congress Control Number: 2020948480

ISBN 978-1-4985-9774-6 (cloth)
ISBN 978-1-4985-9775-3 (electronic)

To Filipe, Beatriz, and Pedro.

Contents

List of Figures and Tables

FIGURES

TABLES

Acknowledgments

The idea of researching public participation in Brazilian EIA flowed as a result of my own experience working as an in-house environmental attorney in the private sector of major companies in Brazil, where I was constantly exposed to the topic: from preparation of Terms of Reference for EIAs to participation and planning public hearings and obtainment of environmental permits for important projects. Therefore, in 2013, I left Brazil to pursue my goal and dedicate some time to dive in this fascinating subject that interconnects democracy, politics, development, and law. The United States, the birthplace of EIA, seemed an appropriate place to carry out my book project. In 2015, I visited Pará, the state of the Volta Grande Project and the Belo Monte Dam, where I carried out interviews with residents and other stakeholders in Senador José Porfírio, Altamira, and Belém.

I presented preliminary version of this book to several audiences including at the International Studies Association fifty-eighth Annual Convention, Baltimore, United States (2017); Association of Transnational Law Schools at Bucerius Law School, Hamburg, Germany (2016); Middle Atlantic Council of Latin American Studies, Philadelphia, United States (2016); the Young Scholars' Conference at the Michigan Law School, Michigan, United States (2015); and at the Mid-Atlantic SJD Roundtable at Georgetown University, Washington, District of Columbia, United States (2015, 2016). These presentations helped me polish some arguments and dive deeper in some aspects of the book. I also coauthored with Dr. Eve Bratman a related article, that was published in 2018, which gave me an ethnographic perspective of participation in Amazonian states.

Several individuals and organizations contributed to the development of this book and I owe my gratitude to each of them. Specially to David Hunter (American University Washington College of Law), whose wise advices and feedback were essential to the development of this work. A huge thank-you to Dr. Eve Bratman (Franklin & Marshall College). I am profoundly grateful for her knowledgeable comments and review. Her perspective on development politic in Brazil and advices from a "Brazilian by heart" definitely enriched my research. William Snape (American University Washington College of Law) for his feedback and challenging questions. The Center of Latin America and Latino Studies at American University, which served as a channel to obtain the Tinker Foundation Field Research Grant. I gained incredible knowledge as well as an extraordinary life experience during the field research. It gave me a sense of the reality of the Volta Grande and Belo Monte projects and it was important to identify the main problems experienced by communities living in Amazonian states.

Kind thanks also to Dr. Eliane Moreira (Ministério Público Estadual do Pará), who first advised me to investigate the Volta Grande Project, provided essential material, and introduced me to some key actors involved in this case. Dr. Moreira's support was of great relevance to carry out this research.

I would like to thank Marcia and Etelvina Penalber for hosting me with care and attention in Belém, in the state of Pará, Brazil. I am also very thankful to Amanda Frost (American University Washington College of Law) who has always listened and encouraged me in many different ways.

Many individuals were extremely kind and played an important role throughout this research. From having an insightful conversation to suggesting academic and legal materials, assisting in planning my visit to Altamira, Belém, and Senador José Porfírio, and connecting me with key stakeholders in the Belo Sun Case, I could not thank enough Luiz Antonio Pazos Moraes (Banco Nacional de Desenvolvimento Econômico e Social), Thaís Santi (Ministério Público Federal—Pará), Dr. Andreia Barreto (Defensoria Pública do Estado do Pará), Mauro Barros (Belo Sun Mining Corporation), Rodrigo Oliveira (Ministério Público Federal—Santarém), João Batista Uchoa Pereira (Fundação Viver, Produzir e Preservar), Jackson Dias (Movimento dos Atingidos por Barragem), Iury Bezerra Paulino (Movimento dos Atingidos por Barragens), Martha Goulart (Neoenergia), Ana Amorim (Ecology and Environment do Brasil), Leonardo Amorim (Université Paris—Panthéon Sorbonne), Biviany Rojas Garzón (Instituto Socioambiental), Raul Silva Telles do Valle (World Wildlife Fund—Brazil), Robert Albro (Center of Latin American & Latino Studies—American University), Marcela Palacio Puerta (Universidad Santo Tomás), Majed Al, Monica Oliveira (Conexão Natureza) and an anonymous reviewer.

My deepest gratitude to the several interviewees who live in Ressaca Village and Altamira, who will remain anonymous. They generously opened their houses for me and agreed to share their views and experiences about the

Belo Monte and Belo Sun projects. Their dedicated time and attention were valuable to the analysis of this book.

I thank my entire family for understanding the difficult decision of leaving Brazil, my country, to pursue this research. Especially my mother, Dagmar Bená Dias, who has always been my inspiration regarding determination, optimism, and persistence. To my father, Jorge Willemenn Dias, for shaping the person who I have become and for his continuous belief in my work. Essentially, publishing this book would not be possible without the encouragement, wise advices, reviews, editions, and endless patience of my husband Filipe Lage de Sousa.

Certainly, this trajectory would be less enjoyable if I did not have Beatriz Dias de Sousa and Pedro Dias de Sousa in my life. Beatriz was a brilliant discussant about public participation and important values a society should have. I continue learning from her desire to make a difference in the world that we live. A huge thank-you to Pedro for his smiles, hugs, kisses, and especially for understanding my absence on weekends during these years. They truly held me up and nourished my passion for learning.

List of Acronyms

BAPE	Bureau the Audiences Publiques sur l'Environnement (Office of Environmental Public Hearings)
BCS	Broad Community Support
BNDES	Banco Nacional de Desenvolvimento Econômico e Social (Brazilian Development Bank)
CAO	Compliance Advisor/Ombudsman
CARU	Comisión Administradora del Río Uruguay (Administrative Committee for Uruguay River)
CE	Categorical Exclusion
CEPRAM	Conselho Estadual de Meio Ambiente—Bahia, Brazil (Environmental State Council—State of Bahia, Brazil)
CEQ	Council on Environmental Quality
CF	Constituição Federal (Federal Constitution)
COEMA	Conselho Estadual do Meio Ambiente—Pará (Estate Environmntal Council—State of Pará, Brazil)
COMPERJ	Complexo Petroquimico do Rio de Janeiro (Rio de Janeiro Petrochemical Complex)
CONAMA	Conselho Nacional do Meio Ambiente (National Environmental Council)
DAD	Decide, Announce, Defend
DEAD	Decide, Educate, Announce and Defend
DINAMA	Dirección Nacional the Medio Ambiente (National Environment Directorate)
DNIT	Departamento Nacional de Infraestrutura de Transporte (National Department of Infrastructure Transportation)

DNPM	Departamento Nacional de Produção Mineral (National Department of Mineral Production)
EA	Environmental Assessment
EC	European Commission
EIA	Environmental Impact Assessment
EIS	Environmental Impact Statement
ESS	Environmental and Social Standard
EU	European Union
FCP	Fundação Cultural Palmares (Palmares Cultural Foundation)
FONSI	Finding of No Significant Impact
FPIC	Free, Prior, and Informed Consent
FUNAI	Fundação Nacional do Índio (National Indian Foundation)
FUNASA	Fundação Nacional de Saúde (Health National Foundation)
GDP	Gross Domestic Product
IACHR	Inter-American Commission on Human Rights
IAIA	International Association for Impact Assessment
IBAMA	Instituto Brasileiro do Meio Ambiente e dos Recursos Naturais Renováveis (Brazilian Institute of Environment and Renewable Natural Resources)
IBRD	International Bank for Reconstruction and Development
ICJ	International Court of Justice
ICMM	International Council on Mining and Metals
IDA	International Development Association
IDB	Inter-American Development Bank
IEMA	Institute of Environmental Management and Assessment
IFC	International Finance Corporation
ILO	International Labor Organization
INCRA	Instituto Nacional de Colonização e Reforma Agrária (National Institute of Agrarian Reform)
IP	Inspection Panel
IPHAN	Instituto do Patrimônio Histórico e Artístico Nacional (National Institute of Historic and Artistic heritage)
MDS	Ministério de Desenvolvimento Agrário (Ministry of Agrarian Development)
MIGA	Multilateral Investment Guarantee Agency
MUM	Meet, Understand, Modify
NCEA	Netherlands Commission for Environmental Assessment
NEPA	National Environmental Policy Act
NGO	Nongovernmental Organization
NPV	Net Present Value
OAS	Organization of American States
OECD	Organization for Economic Cooperation and Development

PAC	Programa de Aceleração do Crescimento (Growth Acceleration Program)
PBA	Plano Básico Ambiental (Basic Environmental Plan)
PLS	Projeto de Lei do Senado (Senate Draft Law)
POP	Public Owns Project
PS	Performance Standard
RIMA	Relatório de Impacto Ambiental (Environmental Impact Report)
SEMA	Secretaria Especial do Meio Ambiente (Special Secretariat for the Environment)
SEMAS	Secretaria de Meio Ambiente e Sustentabilidade—Estado do Pará, Brasil (Secretariat for Environment and Sustainability Estate of Pará, Brazil)
SEPLAN	Secretaria de Planejamento (Planning Secretary)
SISLIC	Sistema Informatizado de Licenciamento Ambiental (Online Environmental Permitting Procedure System)
SISNAMA	Sistema Nacional do Meio Ambiente (National Environmental System)
SJP	Senador José Porfírio (city in the State of Pará, Brazil)
TOR	Terms of Reference
UK	United Kingdom
UNCLOS	United Nations Convention on the Law of the Sea
UNDRIP	United Nations Declaration on the Rights of Indigenous Peoples
UNEP	United Nations Environment Program
USA	United States of America
US EPA	United States Environmental Protection Agency
WB	World Bank

Introduction

Environmental Impact Assessment (EIA) is a procedure that evaluates both the positive and negative effects of proposed projects, programs, or policies.[1] It is an important planning instrument which assists in the integration of environmental consideration in economic decision-making. Therefore, the EIA procedure should guarantee that, prior to approval, (1) decision-makers have fully considered the social and environmental impacts in their decision; and (2) affected communities participated in the process acknowledging the risks and impacts, and were listened by governments.[2] Over the past four decades, international institutions, private sector, and over 150 countries established EIA policies and legislation. Wood states that "EIA is not EIA without consultation and participation."[3] Therefore, it is a fundamental part of the procedure.

Public participation is the involvement of people, normally, in governmental decisions. Levels of public participation are informing, consulting, participating, and negotiating. EIAs around the world require, in general, consultation, which means that stakeholders provide their input and decision-makers should take them into consideration, prior to authorization of projects, programs, or policies.

Social justice aside, public participation is also important in financial terms. Research on the impacts of stakeholders' cooperation or conflicts on a project's budget and schedule demonstrates that "community consent" or "social license" is more than rhetoric. It is a measurable asset, which should be managed during early stages of project planning and EIA procedures.[4] The International Corporate Accountability Roundtable concluded that

> In 1975, tangible assets accounted for up to 80% of the valuation assessment for corporate securities' market value. In 2005, tangible assets accounted for only

> 20% of that valuation assessment, as intangible assets—including risk management, intellectual property, human and social capital—have come to be used to calculate 80% of the market valuation equation for corporations.[5]

Stakeholders' engagement has a direct impact on profit and project's budget estimation. Henisz and Zelner provide the following example: "$15 billion of gold sitting in a mountainside cannot be transformed into shareholder rents with financial, engineering, and marketing inputs alone. It also requires the political and social support of key stakeholders including not only members of the economic value chain but also government officials, regulators, community leaders, and members of civil society."[6] In this vein, a research study analyzing fifty worldwide extractive industry projects demonstrated that the most frequent and major costs faced by extractive industries are those related to temporary shutdown or delays which impact productivity.[7] "For example, a major, world-class mining project with capital expenditure of between US$3-5 billion will suffer costs of roughly US$20 million per week of delayed production in Net Present Value (NPV) terms, largely due to lost sales."[8] The social cost is of course high to those who suffer displacement against their will, water and air pollution, and loss of social and communities' ties.

Although there are good initiatives at the international level—such as the Equator Principles, World Bank Environmental and Social Safeguards, International Finance Corporation (IFC) Performance Standards and IFC's Stakeholder engagement guide—countries, especially developing ones, still face difficulties in including meaningful public consultation within their EIA procedures. Furthermore, in environments plagued by distrust in governments, corruption, and a real need to grow economically, it is a complex task to ensure that decision makers consider their inputs. Consultation, therefore, in some cases, is not enough to ensure that impacts of development will change communities' lives for the better nor will it account their views and expectations. Particularly, in cases where large-scale projects or even medium-sized ones (this last one with significant cumulative impacts) dramatically change their environment, directly affecting their human right to life. Therefore, communities should have a more powerful voice in the authorization of such projects and collaborate in decisions that transform their future.

Free, Prior, and Informed Consent (FPIC) is the right of indigenous and tribal peoples/communities, through a consultation process, to reach consensus or quasi-consensus on the authorization of development projects.[9] This right derives from their self-determination to make decisions upon their future. FPIC does not require a "yes" or "no" answer, yet it encourages a true dialogue among project proponents, governments, and local affected communities on the authorization of projects, mitigation, and compensation measures. EIA is important in conducting such dialogue; however, consultations should happen at early stages and in some cases include a community's education on a project's potential impacts with clarity and transparency. This is certainly a difficult task that requires trust and neutrality. FPIC should be extended to local affected communities whenever projects will cause deleterious impacts in their lives.

This study is carried out within a national context: Brazilian EIA regulations and practices. The 1988 Brazilian Federal Constitution mandates that whenever projects might cause significant environmental impact, an EIA must be developed, and it must be public.[10] Since then, a significant number of EIAs have assessed the environmental feasibility of a variety of activities, ranging from massive infrastructure projects to medium-sized and small potential polluters. As a developing country, which also hosts the largest rainforest in the world, Brazil faces the challenge of equilibrating the equation: growing economically, protecting its natural and cultural resources, and considering its people's participation in governmental decision-making over development projects. Furthermore, decisions regarding the approval of some infrastructure projects have been argued by scholars, the Inter-American Court of Human Rights, and civil society organizations as not truly considering social impacts of implementation of projects and not appropriately carrying out public consultation.[11] Several studies carried out by Brazilian Auditing Court, *Ministério Público*, and the Brazilian Senate point to deficiencies of public participation in Brazilian EIAs. All these studies are discussed throughout the book.

The current Brazilian EIA process falls short on measures and techniques capable of guaranteeing meaningful public participation. This is due to weak regulations, lack of political will, and environmental agencies' deficiencies (such as lack of qualified personnel). Additionally, a project's affected population, especially the poor and less educated, often feel discouraged from providing effective input. There is general disbelief in institutions, especially considering that EIAs are intrinsically marked by bias from those who develop them and from those who suffer political pressure during their technical reviews.

Furthermore, timing is key to ensure effective participation. By the time participation takes place in Brazilian EIAs, chances to influence decision-making are low because financial resources and time have already been significantly invested in the planning and approval of the project. Aside from these factors, other practical aspects contribute to ineffective participation in EIAs: the poor quality of studies, the technical language utilized to present the project and its impacts to communities, and the ineffective means utilized to invite the public to participate, such as publication in newspapers and official gazettes, which do not reach all relevant stakeholders—especially considering that Brazil has 11.3 million illiterate adults.[12] Moreover, public hearings, which should be a forum established for open dialogues, is normally a battle space. Participation in Brazil happens in a reactive fashion.[13] In other words, projects are proposed, and people react. Thus, generally, there is not a fruitful discussion on the projects and their impacts, considering mitigation and compensation measures in a win-win situation.

Brazilian scholars recognize the need to modernize the norms which regulate the EIA.[14] CONAMA (Conselho Nacional do Meio Ambiente) Resolution No. 001 is the main federal regulation on EIA and it dates from 1986. However, concerning public participation, the literature does not set a specific procedure capable of strengthening the development of EIA.[15] This book intends to contribute to the existing knowledge by proposing several steps in the Brazilian federal EIA which moves public participation to earlier stages of this procedure. Therefore, the goal is twofold: to analyze current public participation in Brazilian federal EIA regulation and to propose measures that empower communities to seek their consensus (or not) on authorization of development that will significantly impact their lives.

Therefore, this book seeks to answer the following questions: Who should participate in the EIA and how? How can impartiality in the development of EIAs be ensured? How can the public be empowered to participate effectively? Who should decide on the authorization of projects, and how? A comparative analysis of worldwide best practices, including the United States, international institutions and elsewhere, is applied. Moreover, it analyzes an actual case study: The Volta Grande is a mining project, owned by the Canadian Belo Sun mining corporation, which is planned to be installed in the Amazon rainforest, a few kilometers to the high-profile Belo Monte Dam. This case was chosen due to its peculiarities. There are different groups involved, such as indigenous peoples, artisanal miners, and riverbank communities, who will be impacted by both projects. The case also reveals several of the weaknesses highlighted above as far as public participation in EIA is concerned.

The book is divided in three parts: Part I sets the background providing a description and analysis of international legislation and Brazilian EIA procedure. It describes the main laws and regulations related to the topic as well as the procedure itself. Part II analyzes the case study, the Volta Grande Project, pointing to the main weaknesses of public participation in EIAs and demonstrating how the proposals recommended would reflect in practice. Part III discusses the moment a project is planned and its subsequent engagement in the EIA. It is divided in the EIA procedure stages: screening, scoping, writing the Environmental Impact Statement, public participation, and decision. This classification is used due to its clarity on each step of EIA development and its didactical approach.

NOTES

1. John Glasson et al., *Introduction to Environmental Impact Assessment*, 3rd ed. (London: Routledge, 2012).

2. *See* David Hunter et al., *International Environmental Law and Policy*, 5th ed. (St. Paul: Foundation Press, 2015).

3. C. M. Wood, *Environmental Impact Assessment: A Comparative Review*, 2nd ed. (London: Routledge 2002).

4. Witold J. Henisz et al., "Spinning Gold: The Financial Returns to Stakeholders Engagement," *Strategic Management Journal* 35, no. 12 (December 2013): 1744.

5. International Corporate Accountability Roundtable, *Knowing and Showing: Using U.S. Securities Laws to Compel Human Rights Disclosure* (2013): 29; Pelosi Adamson, *Indigenous Rights Risk Report* (First People Worldwide 2014): 33, http://www.nationalunitygovernment.org/pdf/2014/Indigenous-Rights-Risk-Report-Nov-2014.pdf.

6. Witold J. Henisz and Bennet A. Zelner, "Legitimacy, Interest Group Pressures and Institutional Change in Emergent Institutions: The Case of Foreign Investors and Home Country Governments," *Academy of Management* 30, no. 2 (2005): 361–382.

7. Rachel Davis and Daniel M. Franks, *Costs of Company-Community Conflict in the Extractive Sector* (Cambridge, MA: Harvard Kennedy School, Corporate Social Responsibility Initiative Report No. 66, 2014), https://www.hks.harvard.edu/m-rcbg/CSRI/research/Costs%20of%20Conflict_Davis%20%20Franks.pdf.

8. Ibid.

9. United Nations Declaration on the Rights of Indigenous Peoples (UNDRIP) and the International Labor Organization Convention No. 169 recognize FPIC of indigenous and tribal peoples.

10. Constituição Federal (C.F.) (Constitution) art. 225, §1 IV (Braz.).

11. The Belo Monte Dam is an internationally well-known case, in which the International Court of Human Rights determined that Brazil did not appropriately consult indigenous peoples on the implementation of the dam. Indigenous Communities of the Xingu River Basin, Pará, Brazil, Precautionary Measures, Inter-Am. Comm'n HR., PM 382/10 (2011), http://www.cidh.oas.org/medidas/2011.port.htm.

12. In 2014, Brazil had the eighth highest number of illiterate adults among all countries in the world, according to UNESCO. UNESCO, Teaching and learning: achieving quality for all; EFA global monitoring report, (2013–2014), https://unesdoc.unesco.org/ark:/48223/pf0000225660; *See also* Andre Fuentes, "Brasil é o 8º País Com Mais Adultos Analfabetos do Mundo," *Impávido Colosso* (Undaunted Colossus) (blog), February 12, 2014, http://veja.abril.com.br/blog/impavido-colosso/brasil-e-o-8-pais-com-mais-adultos-analfabetos-do-mundo/.

In June 2019, the Brazilian Institute of Geography published the latest results about illiteracy in Brazil: 11.3 million adults who are over fifteen years old are illiterate in Brazil, https://infograficos.gazetadopovo.com.br/educacao/analfabetismo-no-brasil-em-2018/.

13. Luiz Enrique Sanchez, *Avaliação de Impacto Ambiental Conceitos e Métodos* (Environmental Impact Assessment Concepts and Methodologies), 2nd ed. (São Paulo: Oficina de Textos, 2013).

14. *See* Paulo Affonso Leme Machado, *Direito Ambiental Brasileiro* (*Brazilian Environmental Law*), 21st ed. (São Paulo: Malheiros, 2013); Antonio F. G. Beltrão, *Aspectos Jurídicos do Estudo de Impacto Ambiental (EIA)* (Legal Aspects of Environmental Impact Assessment (EIA)) (São Paulo: MP Editora, 2008).

15. Sanchez proposes some measures such as project proponents' meetings with communities. However, this author thinks that such measures should be under the project proponent's responsibility independent of the existence of legal requirement. Sanchez, *Avaliação de Impacto Ambiental Conceitos e Métodos*.

Part I

EIA IN INTERNATIONAL LAW AND IN BRAZILIAN LEGISLATION

Chapter 1

EIA in International Law

Popular movements against consumption models and development processes appeared worldwide during the 1950s and 1960s. These gave rise to proposals to manage natural resources, particularly with the use of mechanisms of command and control; the development of laws and regulations was one expression of such mechanisms.[1] In addition to this push to put the environment on the development agenda, the Santa Barbara oil spill (in early 1969) was a catalyst, resulting in public pressure on the U.S. government to have mechanisms in place to consider the environment in economic decision-making.[2]

As a result, the United States National Environmental Policy Act 1970 (NEPA) was created. The NEPA was the birthplace of the EIA. Section 102(c) of NEPA requires that "agencies of the Federal Government shall (. . .) include in every recommendation for major Federal actions significantly affecting the quality of the human environment, a detailed statement on (. . .) the environmental impact of the proposed action."[3]

NEPA was an important framework that inspired other countries to adopt EIA legislation, such as Canada in 1973, the Commonwealth of Australia in 1974, New Zealand in 1974, West Germany in 1975,[4] and France in 1976.[5]

EIA REQUIREMENTS IN INTERNATIONAL FINANCIAL INSTITUTIONS' POLICIES

International institutions have adhered to the "think-before-acting" approach by adopting EIA procedures and policies. Thus, in 1974, the Organization for Economic Cooperation and Development (OECD) issued a recommendation advising member governments to "establish procedures and methodologies for forecasting and describing the environmental consequences of significant public and private projects likely to have a major impact on the quality of the environment."[6] In 1985, the Council of the European Communities enacted the EIA Directive (EU EIA Directive or EIA Directive), which was amended in 1997 and entered into force in 1999.[7] The EIA Directive strengthened and fostered the spread of EIA in European countries.

In the early 1990s, international financial institutions were also being pressured by civil society to abandon the classical model of economic growth and foster sustainable growth in developing or at least developed countries. In 1979 the Inter-American Development Bank (IDB) was the first multilateral development bank to adopt an environment operational policy (OP 703). In 1996, the IDB, influenced by the 1992 Rio Declaration, included several mandates in its operations, including quality control and EIAs.[8]

The World Bank (WB or the Bank) adopted Environmental and Social Safeguards, including the Environmental Assessment, as an umbrella policy in 1992. The WB revised its safeguards in 1995 and 1999.[9] In 2012, the Bank began a new process of revision, including several consultations in borrowing countries. On August 4, 2016, its Board of Executive Directors approved new environmental and social framework safeguards.[10]

EIA IN GLOBAL ENVIRONMENTAL INSTRUMENTS

Of the international instruments containing EIA procedures, related principles, and obligations, it is worth mentioning the *Goals and Principles of Environmental Impact Assessment* adopted by the United Nations Environment Program (UNEP) in 1987.[11] These principles provide basic standards, aiming to guarantee sustainable projects and ensuring at least three objectives: environmental impacts should be considered before authorization of projects and activities, national governments should ensure the implementation of EIAs, and states should exchange information and consultation reciprocally regarding projects or activities which have transboundary effects. Currently, the UNEP *Goals and Principles* are widely accepted as state practice and also reproduced in several international instruments.[12] Principle 17 of the 1992 Rio Declaration states that "Environmental Impact Assessment, as a national instrument, shall be undertaken for proposed activities that are likely to have a significant adverse impact on the environment and are subject to a decision of a competent national authority."[13]

The Convention on EIA in a Transboundary Context adopted in Espoo, Finland, in 1991 (the Espoo Convention), which entered into force in 1997, is an important treaty on the subject. The Convention sets forth obligations of parties to carry out EIAs, including notification and consultation among the parties, whenever a major project will have an impact across the borders.[14] Moreover, the 1991 Protocol on Environmental Protection to the Antarctic Treaty requires state parties to cooperate and provide assistance to each other in the preparation of EIAs on activities in the Antarctic Treaty area.[15]

Regional and international treaties on the protection of the seas also have obligations to carry out EIAs whenever the activity might cause environmental degradation, such as the UNEP Regional Seas program,[16] the 1986 Noumea Convention governing the marine and coastal environment of the South Pacific region, and the 1982 United Nations Convention on the Law of the Sea (UNCLOS).[17]

STATUS OF EIA UNDER INTERNATIONAL LAW

During the last forty years, over 150 countries have enacted EIA regulations. Due to the common practice of requiring an EIA whenever a project might cause significant environmental impact, some scholars claim that the duty to carry out an EIA on potentially harmful project impacts has become part of customary law.[18] The two sources of customary law are state practice and *opinio juris*. Baker explains that *opinio juris* means "the consistent practice of states, coupled with the determination (by the practicing state) that such practice is being undertaken out of a sense of legal obligation (labelled *opinio juris*)."[19] It is clear that state practice exists; however, under *opinio juris*, it is not evident whether the duty of states to conduct an EIA results from an international obligation.[20]

Sands states that "the mandatory language of Principle 17 confirms that environmental impact assessments are now required by general international law,"[21] especially in relation to projects which might have transboundary effects. However, Principle 17 does not prescribe further details to orientate states on the requisites of EIAs.[22] According to Sands, the aforementioned nonbinding, binding instruments as well as international cases shed light on the circumstances under which states shall request an EIA prior to the authorization of a project.[23]

In relation to international cases, the International Court of Justice (ICJ) dealt with questions involving the EIA in the New Zealand's application to prevent France from undertaking nuclear testing in the South Pacific (1995);[24] the *Gabcikovo-Nagymaros* case (1997), in which Hungary claimed Slovakia's breach of the Treaty of 16 September (1977) in the construction and operation of the Gabcikovo-Nugymaros System of Locks, when Slovakia constructed a dam on the Danube River;[25] the *Ireland v. United Kingdom* case in relation to the Mox Plant (2001); and the *Uruguay v. Argentina* case concerning the pulp mills on the River Uruguay.[26] In the same vein the International Tribunal on the Law of the Sea issued the advisory opinion on Responsibilities and Obligations in the Area (2011).[27] For the purpose of this book, it is worth looking at the *Pulp Mills* case with a more thorough analysis.

The *Argentina v. Uruguay Pulp Mills* Decision

The importance of the *Pulp Mills* case is twofold: (1) it discusses the consultation and exchange of information among states when the proposed activity might have impacts on shared natural resources and (2) it brings an important statement of the ICJ that EIA is a requirement under general international law. In its 2010 decision the ICJ stated that

> The obligation to protect and preserve, under Article 41(a) and (b) of the 1975 Statute has to be interpreted in accordance to a practice, which in recent years has gained so much acceptance among States that it is considered a requirement under general international law to undertake an environmental impact assessment where there is a risk that the proposed industrial activity may have a significant adverse impact in a transboundary context, in particularly, on a shared resource. Moreover, due diligence and the duty of vigilance and prevention which implies, would not be considered to have been exercised, if a party planning works liable to affect the regime of the river of the quality of its water did not undertake an environmental impact assessment on the potential effects of such works.[28]

In this statement, ICJ affirms that development of an EIA implies the duty of vigilance and prevention, which are important sustainable practices. The obligation to conduct an EIA whenever a project might cause significant transboundary environmental impact exists in international law. Public participation is an important component in this procedure. It is through the EIA that the public can express their views on the proposed projects and contribute to its implementation. However, states' obligations to require public participation as well as how meaningful such participation will happen or how much consideration governments will give to public's input is less clear under international law. In this context, the *Pulp Mill* decision sheds light on the requirement of consultation between states, especially in the context of shared natural resources.

The case addressed the proposed construction of two pulp mills—the CMB (ENCE) and Orion (Botnia)—in the territory of Uruguay near the River Uruguay, which is shared by Argentina and Uruguay.[29] In February 1975, Argentina and Uruguay signed a treaty for the management of the river—the 1975 Statute of the River Uruguay (the 1975 Statute).[30] The Statute created a bilateral commission that would manage the shared use of the river: the

Comisión Administradora del Río Uruguay (CARU).[31] Articles 7 to 12 of the 1975 Statute established that a "party which is planning certain activities" shall notify the CARU, which will then determine whether the project is potentially harmful to the other party and therefore will conduct the dialogue between Argentina and Uruguay.[32]

The court stated that the obligation of consultation (by providing information to CARU) is necessary because it fosters cooperation of the involved parties and therefore fulfills the obligation of prevention.[33] In relation to timing, the court noted that the information should be provided at the preliminary stages of the project, according to Article 7 of the 1975 Statute. In other words, it is not necessary to have a full impact assessment in place to provide information to the other party.[34] However, as more information is acquired, this shall be transmitted to CARU.[35] According to the court, "in any event, the duty to inform CARU will become applicable at the stage when the relevant authority has had the project referred to it with the aim of obtaining initial environmental authorization and before the granting of that authorization."[36]

The court decided that Uruguay did not comply with its obligation to inform CARU on the CMB (ENCE) and Orion (Botnia) mills, although the commission requested information several times. Uruguay sent a summary of the CMB (ENCE) EIA for public disclosure;[37] however, Uruguay did not send any information on Orion (Botnia).

CARU obtained the information regarding the plans to construct the mills through other means (companies and nongovernmental organizations) rather than from the government of Uruguay, which was planning the mills.[38] As the court observed, the role of the EIA is to provide information not only to the planning party but also to the other potentially affected party on the impacts of the project. Therefore, both parties should participate in the EIA in order to consider implementation of the project:

> The Court notes that environmental impact assessments which are necessary to reach a decision on any plan that is liable to cause significant transboundary harm to another state must be notified by the party concerned, through CARU, pursuant to Article 7, second and third paragraphs, of the 1975 statute. This notification is intended to enable the notified party to participate in the process of ensuring that the assessment is complete, so that it can then consider the plan and its effects with a full knowledge of the facts. (Article 8 of the 1975 Statute)[39]

The court observed that notification must happen before authorization is given by the state hosting the project. In this context, the court noted that Uruguay did not inform CARU of its plans to implement the mills and it provided the assessments to Argentina after authorization for the mills was issued. Therefore, the court stated that

> Uruguay ought not, prior to notification, to have issued the initial environmental authorizations and the authorization for construction on the basis of the environmental impact assessments submitted to the *Dirección Nacional the Medio Ambiente*—DINAMA. Indeed, by doing so, Uruguay gave priority to its own legislation over its procedural obligation under the 1975 Statute and disregarded the well-established customary rule reflected in Article 27 of the Vienna Convention on the Law of the Treaties.[40]

The court concluded that Uruguay did not comply with its obligations under Article 7 of the 1975 Statute,[41] as it did not inform, notify, and negotiate with Argentina as required by the Statute.

The *Pulp Mills* decision emphasized the importance of participation from affected parties in the EIA to inform the final decision on approving the project. In this vein, the court stated that notification should have happened before authorization was given from Uruguay. Furthermore, the court noted that the information should have been provided to CARU at the early stages (before an EIA was carried out). Thus, CARU could have decided whether the project would have a significant transboundary impact. And later, according to the court's interpretation, whenever more information was available, the Statute also required the involvement of both parties.

The decision calls attention to the momentum of participation in development plans; it also highlights the importance participation plays in the transboundary context. However, the decision does not touch upon how much participation would be required or how Argentina's views would be considered in Uruguay's final decision on approving the project.

EIA AND ITS MANY DIMENSIONS

The definition of EIA varies. The International Association for Impact Assessment (IAIA) defines it as "the process of identifying, predicting, evaluating and mitigating the bio-physical, social and other relevant effects of development proposals prior to major decisions being taken and commitments made."[42] The Espoo Convention simply states it is "a national procedure for evaluating the likely impact of a proposed activity on the environment."[43] EU EIA Directive 2014/52/EU refers to EIA as a process that consists of the preparation of an environmental impact report by the developer, the consultation of relevant authorities and public participation, and the analysis and decision related to the project in question.[44] Additionally, Munn refers to EIA as the need "to identify and predict the impact on the environment and on man's health and well-being of legislative proposals, policies, programs, projects and operational procedures, and to interpret and communicate information about the impacts."[45]

Although EIA has several definitions, all of them recognize EIA as a preventive process that identifies and predicts environmental and social impacts derived from activities, policies, or programs. Basically, EIA serves two main purposes: (1) it is a tool to aid in the decision-making process as the EIA provides information to ground governmental decisions on whether to approve development or not;[46] and (2) it is an instrument to achieve sustainable development[47] as it predicts impacts and enables the requirement of mitigation measures, or in some cases change the design of projects or even reject projects that will cause too much for people and the environment. Therefore, EIA balances economic, social, and environmental effects of projects or activities.

Furthermore, different actors see EIA in different ways. Some view EIA as merely a political process, which legitimizes governmental decisions that would be taken anyway;[48] others define EIA as a rational, technical, and scientific approach. There are also those who see EIA as a combination of science and political input.[49] Furthermore, for some environmentalists, EIA does not provide full certainty on some environmental impacts; in their view, projects carried out under uncertain circumstances should not be carried out at all.[50] As EIA involves different actors—such as governments, affected parties, project proponents, and other stakeholders—scholars have devoted attention to its communicative and collaborative aspects where governments have the role to mediate conflicting interests over development.[51]

All the EIA "views" appropriately describe its purposes. However, EIA has multifaceted dimensions, which means that it encompasses all those aspects in one procedure. In other words, EIA is a technical procedure necessary to inform governmental decisions. Such decisions have a political component; however, they should take into consideration the impacts pointed out in the EIA as well as stakeholders' inputs.

PUBLIC PARTICIPATION IN EIAs

In the international arena, several instruments deal with the right to participate in environmental decisions, access to information, and access to justice. As early as 1972, Principle 20 of the Stockholm Declaration required "free flow of up-to-date scientific information and transfer of experience must be supported and assisted, to facilitate the solution of environmental problems."[52] The Regional Agreement on Access to Information, Public Participation and Justice in Environmental Matters in Latin America and the Caribbean states that public participation and access to information are crucial to sustainable development.[53]

In most jurisdictions, the level of participation required in EIAs is consultation, which means that stakeholders will be heard; their inputs should be taken into consideration in the decision of approving or rejecting a project. Stakeholders do not have veto powers in relation to planned developments, nor is there any guarantee that their views will be taken into consideration or even influence decision-making. Some scholars defend the idea that indigenous peoples are the only ones who have the right to provide or withdraw consent to development projects affecting their lands—the right of free, prior, and informed consent (FPIC)—however, this view is not unanimous.

FPIC

FPIC means that indigenous peoples give their consent in a manner that is free from coercion—they are fully informed, and they truly understand the project and its impacts on equal footing with governments and project proponents. Furthermore, that consent must be given before the project's authorization is issued.[54] Consensus or consent of indigenous communities shall be obtained according to their customary laws and practices.[55]

FPIC is a fundamental right of self-determination.[56] Indigenous peoples are free to determine their political status and decide upon their development.[57] They have the right to give consent to projects and/or activities that might negatively affect them since their culture and way of living depend on the land they inhabit. Therefore, the impact of projects affects indigenous peoples more heavily than on the rest of the society (communities who live in urban areas for example) due to their ties with nature. In this context, the UN Special Rapporteur on Indigenous Land rights stated that

> The legacy of colonialism is probably most acute in the area of expropriation of indigenous lands, territories and resources for national economic and development interests. In every sector of the globe, indigenous peoples are being impeded in every conceivable way from proceeding with their own forms of development, consistent with their own values, perspectives and interests.[58]

The United Nations Declaration on the Rights of Indigenous Peoples (UNDRIP) clearly recognizes FPIC.[59] Article 10 provides that "Indigenous peoples shall not be forcibly removed from their lands and territories. No relocation shall take place without free, prior and informed consent of the indigenous peoples concerned and after agreement on just and fair compensation."[60] Article 19 says that "States shall consult and cooperate in good faith with the indigenous peoples concerned through their own representative institutions in order to obtain their free, prior and informed consent before adopting and implementing legislative or administrative measures that may affect them."[61] Finally, Article 32 provides that

> States shall consult and cooperate in good faith with the indigenous peoples concerned . . . in order to obtain their free, prior and informed consent prior to the approval of any project affecting their lands or territories and other resources, particularly in connection with the development, utilization or exploitation of mineral, water or other natural resources.[62]

The UNDRIP is not a binding instrument;[63] however, according to Mackay, it "should not be merely referred as soft law or aspirational" because it restates several international binding requirements.[64] The International Labor Organization (ILO) Convention No. 169 provides for FPIC in case of displacement.[65] Although the ILO Convention No. 169 is legally binding, the fact that it was ratified by only twenty-two countries raises concerns on its legal significance in the international arena.[66]

The Inter-American Commission on Human Rights (IACHR) has affirmed the right of indigenous communities to FPIC whenever their territory will be potentially affected by a project. For instance, in the *Awas Tingni* case, the IACHR held that Nicaragua violated rights to property since it did not obtain the Awas Tingni community's consent to grant logging concessions on their land.[67] Also, in the *Maya Indigenous communities* case, the IACHR stated that states have the obligation to seek indigenous peoples' consent in order to grant authorization for activities that will impact their land and their communities, such as granting concessions to exploit natural resources within indigenous lands.[68] In March 2007, the IACHR made a very important statement in relation to FPIC, in the *Saramaka People* case, holding that "agents of the State itself, or third parties acting with its acquiescence or its tolerance, to affect the existence, value, use or enjoyment of the territory to which the members of the Saramaka people are entitled, unless the State obtains the free, informed and prior consent of the Saramaka people."[69]

The means and the time when information is provided are crucial in achieving consent. Goodland states that mere explanation of the project and its impacts might not be enough to fully inform indigenous peoples.[70] It is impossible to agree on a mining project or a reservoir when these communities have never seen such projects. Thus, it would be helpful to take indigenous communities and/or their representatives to visit similar projects and talk to people who live in the surrounding areas.[71]

Although it is clear that UNDRIP requires FPIC prior to a project's authorization, it is not so evident in practice what the term "consent" involves.[72] Some practitioners and scholars define consent as the right to say yes or no to projects.[73] However, according to the ILO Convention No. 169, consent does not mean veto power.[74]

The importance of FPIC was recognized in the International Council on Mining and Metals (ICMM), in which twenty-two companies in the mining and metals sector as well as thirty-five associations compromised themselves to adhere to the principles of FPIC in their activities. Despite the importance of ICMM in the industry practice, it does not guarantee indigenous peoples veto powers. In other words, governments can still authorize projects despite indigenous peoples' objection.[75]

Sauer et al. point out that in countries where FPIC is incorporated into national law, such as Peru and the Philippines, companies' financial and reputational risks are minimized.[76] Conversely, in Canada, where national legislation does not incorporate FPIC, governments do not consider consent of indigenous peoples as a right to reject development. Therefore, business risks are maximized. In this context, a 2013 Report of the First Peoples Worldwide points out that in Canada, Southwestern invested US$28.2 million in a project in New Brunswick; however, blockage and opposition of the Elsipogtog and Mi'kmaq First Nations resulted in daily losses of US$60,000.[77]

There is still no common agreement on the application of FPIC as a right to reach consensus between indigenous communities and governments for approving projects. Furthermore, the constant policies' adoption of the word "consultation" rather than "consent" might give the impression that mere consultation of indigenous communities is legitimate and enough. Consent or consensus does not require agreement by all; Goodland's interpretation is that having the majority consensus or consent would suffice to approve or deny a development project.[78] In some cases referendums are used to consult indigenous communities in relation to development projects that would potentially affect their lands, in an effort to achieve consent.[79]

International Finance Institutions' Provisions on Consultation and Consent

Consultation and consent are means of public participation. Meaningful participation may lead to consent.[80] Normally, EIAs include some kind of consultation (public hearings, written comments, scope meetings, among others), which should be taken into consideration by decision-makers.

The former WB Operational Policy 4.01 required one or more forms of environmental assessment, including an EIA. Paragraph 14 stated that for certain projects borrower countries must consult affected groups and local nongovernmental organizations on the proposed project and its environmental aspects. Such consultation should happen as early as possible and public views should be taken into account.[81] It also stated that in order to achieve meaningful consultation between affected groups and nongovernmental organizations, information must be provided in a timely manner and in accessible language.[82] However, the operational policy did not clarify what meaningful participation was nor did it define what modes of participation borrower countries should adopt. To address these and other weaknesses, the WB revised its safeguards and adopted a new Environmental and Social Framework in 2016. In this framework the Environmental and Social Standard No. 10 (WB ESS 10) defines meaningful consultation, among other things, as a two-way process that begins at early stages of the project, and is transparent and free from external manipulation.[83]

Regarding indigenous communities, the WB ESS 7 recognizes that indigenous peoples may be vulnerable to impact of projects in their land and requires the borrower to obtain their free, prior, and informed consent when the project impacts land and natural resources subject to traditional ownership or under customary use or occupation or when it causes their relocation or significantly impacts their land.[84] ESS 7 provides an important provision that guarantees indigenous peoples the right to be consulted and the obligation of the borrower to obtain consent. Paragraph 27 states that when the consent of indigenous peoples cannot be obtained, that specific part of the project will not be processed further. ESS 7 states that consent is the collective support of affected indigenous peoples reached through a culturally appropriate process. The consent might exist even if some individuals or groups are against the project.[85]

The 2012 Sustainability Policy of International Finance Corporation (IFC), which is the private branch of the WB Group, determines that whenever a project is likely to generate potential significant adverse impacts on communities or indigenous peoples, IFC expects its clients to engage in a process of informed consultation and participation. Such procedure will lead to Broad Community Support (BCS) for the project by affected communities. IFC, through its own criteria, will determine whether "the client's community engagement is one that involves informed consultation and participation and enables the participation of the Affected Communities."[86] The Sustainability Policy clarifies that "Broad Community Support (BCS) is a collection of expressions by Affected Communities, through individuals or their recognized representatives, in support of the proposed business activity.[87] There may be BCS even if some individuals or groups object to the business activity."[88] However, it is not clear what BCS is in practice and how IFC establishes whether BCS is present.

The IFC Performance Standard 7 (PS 7) requires clients who intend to develop a project that will adversely affect indigenous communities to undertake an FPIC process, where time should be provided to indigenous peoples' decision-making processes.[89] Paragraph 15 states that "The client will consider feasible alternative project designs to avoid the relocation of Indigenous Peoples from communally held lands and natural resources subject to traditional ownership or under customary use. If such relocation is unavoidable the client will not proceed with the project unless FPIC has been obtained as described above."[90]

The IDB Environment and Safeguards Compliance Policy requires borrowers to carry out an EIA for certain operations.[91] These operations require consultations with affected parties and consideration of their views. Notably, the Operational Policy on Indigenous People merely states that consultation should be in a manner appropriate to the circumstances, with a view to reaching agreement or obtaining consent.[92] Therefore, in practice, who should be consulted and who should provide consent to authorize development projects is not well-defined.

Who Should Be Consulted and Who Should Provide Consent?

There is a misconception and underestimation of the term "consent." Goodland states that FPIC should be given to all affected communities and not only to indigenous peoples.[93] Although indigenous peoples are inherently the owners of the land they occupy, which they depend upon physically and culturally, communities affected by massive projects should also have the right to provide or withhold consent on projects that will significantly affect their lives. The idea is not to undermine indigenous peoples' rights but to reinforce and enlarge such right to other groups who are also heavily impacted by certain projects, as well as the indigenous peoples who are dependent on the land for subsistence.

Practice has demonstrated that in different countries, riverine peasants, the poor and rural populations, have been suffering unequally with development in comparison with others. This happens for many reasons: (1) because normally these people live in distant places where certain development projects are attractive; (2) because communication on the impacts of projects is not effective, therefore they cannot participate meaningfully in EIA procedures; (3) because the local population normally does not have any political influence capable of stopping destructive projects; (4) because unbalanced powers, whenever a project is discussed, sometimes lead to communities feeling intimidated in a public meeting impairing their participation; and (5) because governments do not take into consideration the views of the local population since they occupy land for which they do not have titles or because it is convenient just to remove slum, riverbank, or rural communities.[94] In this context Goodland poses a very provocative but important question: "Why is it that the rural poor can be displaced against their will, but other people cannot? Can development have a double standard and advocate democracy for some, but autocracy for the rest?"[95]

Questions that normally arise are that if local affected communities have FPIC rights, will a minority have such strong power in their hands? In cases where national interest is at stake, will the voice of a small group override the interest of one entire country? For instance, energy is a necessity, yet it is well-known that hydraulic power plants have serious social impacts. Will the locally affected community have the power to stop a project that would benefit the country with energy? There are no fundamental rights more valuable than others. Individuals of a country have the human rights to development in the same fashion that local communities should have the right to decide upon their future, which in many cases mean their right to life.[96] Moreover, due to corruption in many countries, certain projects are authorized and developed, yet benefits to the region or even to the nation never happen. In order to allow consent or FPIC from local communities, certain explicit criteria should be evaluated and applied.

CONCLUSION

The U.S. NEPA was the first national norm to adopt EIA whenever a project significantly affects the environment. It was followed by several states' regulations and multilateral institutions. Soft laws such as the UNEP Goals and Principles of EIA and Principle 17 of the Rio Declaration, as well as binding instruments such as the 1991 Espoo Convention and the 1982 United Nation Convention on the Law of the Sea, reinforce the necessity of carrying out EIAs.

Some scholars view EIA as customary international law; however, this remains arguable since there is no evidence that state practice of carrying out EIAs results from international obligations. The ICJ made an important statement in this direction in the *Pulp Mills* decision, stating that EIA is a requirement under general or customary international law at least when shared natural resources are impacted. Furthermore, the court called attention to the importance of consultation and exchange of information among countries that share natural resources and whenever a project might impact such resources.

While international law is clear that an EIA is required for projects that might cause transboundary impacts, it is not so evident (neither in national or international practices) when public participation occurs in this procedure or how much weight the public's input carries in the final decision-making for these development projects. In most jurisdictions, the level of participation required in EIAs is consultation. More broadly, consultation means the right to be heard, where governments should take public views into consideration during decision-making. However, in practice, there is no guarantee that this will happen in the authorization of projects.

Indigenous peoples have the right to FPIC recognized in UNDRIP and ILO Convention No. 169. However, in practice they have been struggling to have this right recognized and respected, especially since there is not a common definition of FPIC rights on development institutions' policies and national laws. Yet, decisions of the IACHR have determined that indigenous peoples and communities must provide consent prior to authorization of projects. While some scholars and practitioners understand that consent means veto powers, the International Labor Organization has stated that under ILO Convention No. 169 consent does not mean veto power; however, decisions must not be imposed on indigenous communities. In other words, for the ILO, consent is a process where meaningful dialogue leads to consensus.

FPIC is an important tool that guarantees affected communities' right to life, especially considering that many large-scale projects directly affect their land, livelihood, and quality of life. Because of the impact of these projects, FPIC should be extended to other local affected communities and not to indigenous peoples alone. Under certain circumstances, consultation is not enough to protect affected communities from deleterious impacts of development. Building on the idea of consent as a process of meaningful public participation, this book proposes a series of steps to incorporate access to information and public consultation at early stages of EIA in Brazil to achieve consensual decisions on implementation of projects

NOTES

1. Aline Borges do Carmo and Alessandro Soares da Silva, "Licenciamento Ambiental Federal no Brasil: Perspectiva Histórica, Poder e Tomada de Decisão em um Campo em Tensão" (Federal Environmental Permitting in Brazil: Historical Perspective, Power and Decision-making in a Field Experiencing Tension), *Revista Franco-Brasileira de Geografia* 19 (2013): 7, 8.

2. UNEP, "Training Manual," (Environmental Impact Assessment Review Training Workshop, Kingstown, June 7–11, 2011), http://cep.unep.org/iwcam/documents/meeting-reports/environmental-impact-assessment-review-training-workshop; Jones Carys et al., "Environmental Assessment: Dominant or Dormant," in *Taking Stock of Environmental Assessment: Law, Policy and Practice*, eds. Jane Holder and Donald BcGillivray, 3rd ed. (London: Routledge-Cavendish, 2007), 17–19. The Santa Barbara canal in California received more than three million gallons of oil due to a "blow out" from an oil platform. *See* Miles Corwin, "The Oil Spill Heard 'Round the Country," *L.A. Times*, January 28, 1989, http://articles.latimes.com/1989-01-28/news/mn-1162_1_oil-spill/2.

3. National Environmental Policy Act of 1969, Pub. L. No. 91-190, §102 (c), 83 Stat. 852 (codified at 42 U.S.C. § 4321 [2012]).

4. John Glasson et al., *Introduction to Environmental Impact Assessment*, 3rd ed. (London: Routledge, 2012), 36–37.

5. Carys et al., "Environmental Assessment," 17, 18, 19.

6. OECD, "Recommendation of the Council on the Analysis of the Environmental Consequences of Significant Public and Private Projects," C (74)216, November 14, 1974.

7. "Environmental Impact Assessment-EIA," European Commission, accessed September 17, 2015, http://ec.europa.eu/environment/eia/eia-legalcontext.htm.

8. Inter-American Development Bank, *Environment and Safeguards Compliance Policy*, January 19, 2006, http://idbdocs.iadb.org/wsdocs/getdocument.aspx?docnum=665902.

9. Glasson et al., *Introduction to Environmental Impact Assessment*, 37.

10. "Review and Update of the World Bank Safeguard Policies," World Bank, accessed September 4, 2016, http://consultations.worldbank.org/consultation/review-and-update-world-bank-safeguard-policies.

11. UNEP Governing Council, Goals and Principles of Environmental Impact Assessment: Decision 14/25, UNEP/GC/DEC/14/25 (January 16, 1987).

12. Ibid.

13. UN Conference on Environment and Development, Rio Declaration on Environment and Development, Principle 17, A/CONF.151/26/Rev.1 (Vol. I), Annex I (Aug. 12, 1992).

14. Convention on Environmental Impact Assessment in a Transboundary Context, Article 1.VI, Feb. 25, 1991, 309 U.N.T.S. 1989.

15. Protocol on Environmental Protection to the Antarctic Treaty, arts. 6, 8, Oct. 4, 1991, 30 I.L.M. 1461.

16. "Regional Seas Program 1974," UN Environmental Project, last accessed November 16, 2019, https://www.unenvironment.org/explore-topics/oceans-seas/what-we-do/working-regional-seas/regional-seas-programmes/un-environment.

17. Convention for the Protection of the Natural Resources and Environment of the South Pacific Region (Noumea Convention), Nov. 25, 1986, 26 I.L.M. 38 (1987); United Nations Convention on the Law of the Sea, Dec. 10, 1982, 1833 U.N.T.S. 397.

18. David Hunter et al., *International Environmental Law and Policy*, 5th ed. (St. Paul: Foundation Press, 2015).

19. Roozbeh B. Baker, "Customary International Law in the 21st Century: Old Challenges and New Debate," *European Journal of International Law* 21, no. 1 (2010):173.

20. Hunter, *International Environmental Law and Policy.*

21. *See generally* Grigory Tunkin, "Is General International Law Customary Law Only?," *European Journal of International Law* 4, no. 4 (1993): 534.

22. Philippe Sands et al., *Principles of International Environmental Law*, 3rd ed. (Cambridge: Cambridge University Press, 2012), 606.

23. Ibid.

24. Nuclear Tests (N.Z. v. Fr.), Judgment, 1974 I.C.J. 457 (Dec. 20, 1974).

25. Gabcikovo-Nagymaros Project (Hung. v. Slovk.), Judgment, 1997 I.C.J. 3 (Feb. 5).

26. Mox Plant (*Ir. v. U.K.*), ITLOS Case No. 10, Order of Dec. 3, 2001, https://www.itlos.org/fileadmin/itlos/documents/cases/case_no_10/Order.03.12.01.E.pdf; Pulp Mills on the River Uruguay (*Arg. v. Uru.*), Judgment, 2010 I.C.J. 135 (Apr. 20).

27. Responsibilities and Obligations of States Sponsoring Persons and Entities with Respect to Activities in the Area, ITLOS Case No. 17, Advisory Opinion of Feb. 1, 2011, 10 ITLOS Rep. 10.

28. *Pulp Mills*, 2010 I.C.J. 135.

29. Ibid.

30. Ibid., para. 1.

31. Ibid., para. 86.

32. Ibid., para. 50.

33. Ibid., para. 102.

34. Ibid., para. 105.

35. Ibid.

36. Ibid.

37. Ibid., para. 106.

38. Ibid.

39. Ibid.

40. Ibid., para. 121. Article 27 of the Vienna Convention on the Law of the Treaties states that "[a] party may not invoke the provisions of its internal law as justification for its failure to perform a treaty." Vienna Convention on the Law of Treaties art. 27, May 23, 1969, 1155 U.N.T.S. 331.

41. Ibid., para. 122.

42. International Association for Impact Assessment, *What is Impact Assessment?*, Oct. 2009, https://www.iaia.org/uploads/pdf/What_is_IA_web.pdf.

43. Convention on Environmental Impact Assessment in a Transboundary Context, art. 1. (VI), Feb. 25, 1991, 309 U.N.T.S. 1997.

44. Directive 2014/52/EU of the European Parliament and of the Council of 16 April 2014 Amending Directive 2011/92/EU on the Assessment of the Effects of Certain Public and Private Projects on the Environment, 2014 O.J. (L 124) 1.

45. Glasson et al., *Introduction to Environmental Impact Assessment*, 3.

46. Paulo Cesar Gonçalves Egler, "Improving the Environmental Impact Assessment Process in Brazil" (PhD diss., School of Environmental Sciences University of East Anglia, 1998), 63, http://busca.ibict.br/SearchBDTD/search.do?command=search&q=+instituicao_defesa:%22University%20of%20East%20Anglia%22.

47. Glasson et al., *Introduction to Environmental Impact Assessment*, 8.

48. Robert B. Beattie, "Everything you Already Know about EIA (But Don't Often Admit)," *Environmental Impact Assessment Review* 15, no. 2 (1995): 109.

49. Glasson et al., *Introduction to Environmental Impact Assessment*, 14; Egler, "Improving the Environmental Impact Assessment Process in Brazil," 63–64.

50. Glasson et al., *Introduction to Environmental Impact Assessment*, 13.

51. Ibid., 14.

52. Sands, *Principles of International Environmental Law*, 698.

53. Regional Agreement on Access to Information, Public Participation and Justice in Environmental Matters in Latin America and the Caribbean, March 4, 2018, https://repositorio.cepal.org/bitstream/handle/11362/43583/1/S1800428_en.pdf.

54. Robert Goodland, "Free, Prior and Informed Consent and the World Bank Group," *Sustainable Development Law & Policy* 4, no. 2 (2004): 66.

55. Fergus Mackay, "Indigenous Peoples and International Financial Institutions," in *International Financial Institutions and International Law*, eds. Daniel D. Bradlow and David B. Hunter (Austin: Wolters Kluwer, 2010), 287, 303.

56. The right to self-determination is recognized in the International Covenant on Economic, Social and Cultural Rights. International Covenant on Economic, Social and Cultural Rights, art. 1, Dec. 16, 1966, 993 U.N.T.S. 3.

57. United Nations Declaration on the Rights of Indigenous Peoples, art. 3, A/RES/61/295 (Sept. 13, 2007).

58. Erica-Irene A. Daes, Special Rapporteur, "Indigenous People and their Relationship to Land," UN. Doc. E/ECN. 4/Sub.2/2001/21, 49–50 (June 11, 2001).

59. Ipshita Chaturvedi, "A Critical Study of Free, Prior and Informed Consent in the Context of the Right to Development—Can Consent be Withheld?" *Journal of Indian Law & Society* 5 (Winter 2014): 37.

60. United Nations Declaration on the Rights of Indigenous Peoples, art. 10.

61. Ibid., art. 19.

62. Ibid., art. 32.

63. Ipshita Chaturvedi, "A Critical Study of Free, Prior and Informed Consent in the Context of the Right to Development—Can Consent be Withheld?."

64. Mackay, "Indigenous Peoples and International Financial Institutions," 287, 317.

65. Goodland, "Free, Prior and Informed Consent and the World Bank Group," 66.

66. Mauro Barelli, "Free, Prior and Informed Consent in the Aftermath of the UN Declaration on the Rights of Indigenous Peoples: Developments and Challenges Ahead," *International Journal of Human Rights* 16, no. 1 (2012): 5.

67. *The Mayagna (Sumo) Awas Tingni Community v. Nicaragua*, Merits, Reparations and Costs, Judgment, Inter-Am. Ct. H.R. (ser. C) No. 66, 25 (Aug. 2001) citing *The Mayagna (Sumo) Awas Tingni Community v. Nicaragua*, Case 11.577, Inter-Am. Comm'n H.R., Report No. 27/98, 142 (Mar. 3, 1998).

68. *Maya Indigenous Communities of the Toledo District v. Belize*, Case 12.053, Inter-Am. Comm'n. H.R., Report No. 40/04, Merits, 142 (Oct. 12, 2004).

69. *Twelve Saramaka Clans (Lös) v. Suriname*, Case 12.338, Inter-Am. Comm'n. H.R., Report No. 09/06, Admissibility and Merits, 214 (Mar. 2, 2006).

70. Goodland, "Free, Prior and Informed Consent and the World Bank Group," 66.

71. Ibid.

72. Luke Sauer et al., *Consent within Consultation. Incorporating New Business Practices in the Extraction Industry*, CIGI Graduate Fellows Series 3, 1 (2015).

73. Ibid.

74. Brigitte Feiring, *Understanding the Indigenous and Tribal Peoples Convention, 1989 (No. 169)* (International Labor Organization, 2013), https://www.ilo.org/wcmsp5/groups/public/---ed_norm/---normes/documents/publication/wcms_205225.pdf.

75. Sauer, *Consent within Consultation*, 1.
76. Ibid.
77. Ibid.
78. Goodland, "Free, Prior and Informed Consent and the World Bank Group," 66.
79. Brant MacGee, "Participation with a Punch: Community Referenda on Dam Projects and the Right to Free, Prior, and Informed Consent to Development," *Water Alternatives* 3, no. 2 (2010): 162.
80. Goodland, "Free, Prior and Informed Consent and the World Bank Group," 66.
81. The World Bank, *Operational Manual*, Operational Policy 4.01 (January 1999, revised April 2013), https://policies.worldbank.org/sites/ppf3/PPFDocuments/090224b0822f7384.pdf.
82. Ibid.
83. The World Bank, *Environmental and Social Framework*, 99–100 § 22.
84. Ibid., 79–80, § 24.
85. Ibid., 79–80, § 26.
86. International Finance Corporation, *Policy on Environmental and Social Sustainability*, January 1, 2012, http://www.ifc.org/wps/wcm/connect/7540778049a792dcb87efaa8c6a8312a/SP_English_2012.pdf?MOD=AJPERES.
87. Ibid.
88. Ibid.
89. International Finance Corporation, *Performance Standard 7*, 2012, http://www.ifc.org/wps/wcm/connect/1ee7038049a79139b845faa8c6a8312a/PS7_English_2012.pdf?MOD=AJPERES.
90. Ibid., § 15.
91. Inter-American Development Bank, *Environment and Safeguards Compliance Policy*.
92. Inter-American Development Bank, *Operational Policy on Indigenous Peoples and Strategy for Indigenous Development*, February 22, 2006, http://idbdocs.iadb.org/wsdocs/getdocument.aspx?docnum=2032081.
93. Goodland, "Free, Prior and Informed Consent and the World Bank Group," 66.
94. *See* Kristin Sharon, *Environmental Justice: Creating Equality, Reclaiming Democracy* (Oxford: OUP, 2002).
95. Goodland, "Free, Prior and Informed Consent and the World Bank Group," 66.
96. *See* Alexandre de Castro Coura and Lívia Avance Rocha, "Tutela Jurisdicional do Meio Ambiente, e Fenômeno da Colisão de Direitos Fundamentais na Jurisprudência do STF e do STJ" ("Jurisdictional Protection of the Environment, and Phenomenon of Collision of Fundamental Rights in the Jurisprudence of the Supreme Court and the Superior Tribunal of Justice"), *Revista de Direito Ambiental* 19, no. 76 (2014): 37–60.

Chapter 2

EIA in Brazil

SOCIAL, ECONOMIC, AND POLITICAL CONTEXT

This section provides a brief context of Brazilian reality during the emergence of EIA and subsequent developments (from 1964 to 2002). While it is not intended to give a deep social, economic, or political analysis of events, it aims to give the reader a flavor of occurrences that finally led Brazil to create mechanisms to prevent environmental harms or at least to influence decision-making for authorization of projects. Additionally, it is also interesting to see the evolution of thought, along the years, regarding economic growth and environmental protection in a parallel with the conquest of democracy. The section is organized in chronological order.

From 1964 to 1984, the Armed Forces were in power in Brazil establishing an authoritarian regime. During this period most of the policies, including environmental ones, were developed. This was the result of lack of structured environmental norms and international pressure. The goal of the military government was twofold: economic development and international prominence; these would be achieved through the Doctrine of National Security, which was based on "progress and development." The National Security Doctrine aimed to integrate the national territory, modernize the national economy based on the tendency of capitalist nations, such as the United States, however maintaining its independency.[1] During the period of the doctrine, the economy grew so rapidly that this was known as the Brazilian economic miracle. There was an expansion of investments in several sectors, such as industrial and agricultural, as well the construction of hydro dams, roads, and other infrastructure projects reached its peak. The Itaipu dam and the Trans-Amazonian Highway are examples of projects implemented during this period of economic ascension. Brazil's authoritarian regime was eager to grow, therefore, there was no environmental preoccupation during the implementation of these projects. Although there were more jobs in the civil construction, it did not mean increase in salaries for nonspecialized workers. The government did not authorize companies to raise workers' salaries unless the price of final products was unaffected. The media (newspaper, magazines, TV, radio) was censored; otherwise, these means of communication could put the people against the government and prejudice the economic growth of the country and, ultimately, the dictatorial regime.[2]

The miraculous economic growth did not last very long. After the oil crisis in the 1970s, the Brazilian economy was reshaped. There was an increase in the social gap between the poor and the rich. Additionally, the living standard deteriorated significantly, where the lower classes suffered with deficiencies in sewage provision, water supply, and health services.[3]

In 1972, the Stockholm Conference was held, giving more visibility to environmental movements and nongovernmental organizations (NGOs) in Brazil.[4] Representatives of the United Nations as well as several environmental NGOs from all over the world attended the conference. Along with other developing countries, the values that Brazil expressed at that time were based on protecting its own interests. During the Stockholm Conference, Brazil defended the following ideas:[5]

- Economic growth would be a way to prevent environmental pollution in poor areas of the country.
- National development could not be prevented when international recommendations or commitments were not based on robust scientific consensus.
- Developed countries should bear the cost of mitigating environmental deterioration at the international level.

Brazil presented a defensive position in the conference, stating that it would not allow any interference in its sovereignty. At that time, fire in the Amazon rainforest was an international concern and Brazil was worried about international interference in the rainforest.[6] Brazil took the lead among the southern countries in promoting the idea that the discourse on environmental protection during the conference was an excuse to slow down the economic growth of underdeveloped countries.[7]

After the Stockholm Conference, Brazil created a Special Secretariat of the Environment (in Portuguese *Secretaria Especial de Meio Ambiente* [SEMA]), which was functioning under the Ministry of the Interior. SEMA's role was limited to controlling environmental pollution, primarily related to water resources. The idea was merely to control pollution and continue developing economically, without having the obligation to prevent environmental damages. The structure of SEMA was weak—its staff was insufficient, and the budget was low.[8] This was a discrepancy compared to the resources of other secretariats and ministries, which were much more powerful, such as the Secretariat of Planning (in Portuguese *Secretaria de Planejamento* [SEPLAN]). SEMA was terminated in 1989 and the Brazilian Institute of Environment and Renewable Natural Resources was created (in Portuguese *Instituto Brasileiro do Meio Ambiente e dos Recursos Naturais Renováveis* [IBAMA]).

At the end of 1970s, the state had great control over the economy—it was responsible for 40 percent of the country's gross capital formation and the number of federal companies rose from 79 to 131 companies between 1970 and 1976.[9] The change from ministries to secretariats, which responded directly to the president, was a way to control different areas of the executive branch.[10]

During the early 1980s, inflation was very high in Brazil. It was a decade of severe economic crisis and international debts. As a result of international institution requirements (mainly the International Monetary Fund and the World Bank), Brazil had to cut expenses in several areas of the government, including the environmental area. However, at the end of the decade, Brazil recovered and became the eighth largest economy in the world. Industrial, agricultural, and other sectors were in expansion. However, income distribution was still very disproportional. One important advance of this period was the dissemination of NGOs, especially environmental ones.[11]

In the political arena, a great movement to regain democracy started to grow. The *Diretas Já* (direct vote now) campaign was launched by opposition parties and other groups. In several states of Brazil, a great number of people were united in the *Diretas Já* campaign to obtain the right to direct vote. For instance, in São Paulo, around 1,000 people participated in the *Diretas Já*. This was an important movement that through the voice of the people the Congress changed the voting system; however, it was only in 1989 that Congress approved the direct vote for presidential elections.[12] The 1980s faced big changes in the environmental arena. During this period, several laws and regulations were enacted, including the National Environmental Policy and the regulation about environmental permitting and EIA procedures.

As a result of aggressive economic growth, during late 1970s and early 1980s, pollution was increasing exponentially, which called the attention of international and national media. For instance, the city of Cubatão, in São Paulo, reached a level of air pollution that was much higher than what was healthy and acceptable. To minimize this situation, some measures were implemented to prevent pollution; although still timid, mechanisms of prevention started surging. This was done with the enactment of regulation to establish zoning areas and EIAs were being conducted.[13] EIAs were initially carried out even without an existing regulation, which was not enacted until 1980.

The creation of EIAs was a result of endogenous and exogenous factors. There was external pressure to adopt a structured system to evaluate impacts of projects and influence decision-making. While international financial and multilateral institutions played a role, academics were already carrying out studies about the negative impacts of massive projects, such as studies on the dams on the Tietê river, in São Paulo. Internal pressure was exercised also by social groups—such as the Movimento dos Atingidos por Barragens and other environmental groups—which opposed the government's infrastructure projects during the 1970s and 1980s and its lack of environmental concerns during the construction and implementation of these projects as a result of the development model adopted by Brazil.[14]

The year 1992 was emblematic to the environment in Brazil. The Conference of the United Nations on the Environment (Rio-92) was held in Rio de Janeiro, Brazil. The conference influenced a change in ideologies in relation to environmental matters in Brazil, switching from the idea of pollution control to an economic developed based on socio-environmental matters, focusing especially on the idea that the environment and social issues are intimately related. NGOs became participative in policies, including environmental ones.[15]

After 1998, following several economic crises, Brazil started growing economically and encouraging social improvements. The population's purchasing power was stronger. In 2002, there were several infrastructure projects under permitting procedures. Although IBAMA increased its staff in 2002, lengthy permitting procedures persisted.[16] In sum, Brazil faced economic, political, and social challenges. Democracy was regained with effort although Brazil is still in the process of becoming a full democracy. With all the turbulence of the 1970s and 1980s, including discrepant social gaps and the fight to regain democracy, Brazil established its environmental institutions and enacted the related regulations. While not perfect, these initiatives are definitely an incredible step toward sustainable economic growth.

LEGAL HISTORY AND ENVIRONMENTAL STRUCTURE

This section describes the evolution of EIA in Brazilian legislation. It demonstrates how the understanding of EIA, its requirements, and the competence to review EIA has changed over time. The main objective is to provide a brief legal background, presented in chronological order, that points to some shortcomings in the legislation regarding EIAs.

The EIA first appeared as a legal requirement within the scope of the Law No. 6803, of February 7, 1980, which established Basic Guidelines for Industrial Zoning in Critically Polluted Areas. This was the first legislation to require environmental impact evaluations prior to the approval of an activity.[17] Article 10, Paragraph 3 states that the approval of areas used for industrial purposes only will be preceded by special studies containing alternatives and impact assessment, which will establish the reliability of the alternative to be adopted. As the name already suggests, the EIA requirement in this law was limited to only two situations: zones of industrial purpose and critically polluted areas.[18]

One year later, the National Environmental Policy was implemented through the enactment of the Law No. 6938, of August 31, 1981. Article 9 sets forth several tools to assist its implementation, including the official inclusion of EIAs and a permitting procedure. It also established the environmental governance in Brazil. Table 2.1 summarizes the Brazilian environmental structure and its responsibilities, which are relevant to EIAs.[19]

The National Environmental Policy was regulated by Decree No. 88.351, of June 6, 1983, which was later replaced by Decree No. 99.274, of June 6, 1990. These decrees established that the Environmental Impact Evaluation was required within the environmental permitting procedure. The National Environmental Council (in Portuguese *Conselho Nacional do Meio Ambiente* or CONAMA) is responsible for setting up basic criteria of environmental impact evaluation and environmental permitting requirements.[20]

CONAMA regulates Environmental Impact Evaluation through the release of resolutions. CONAMA Resolution No. 001, of January 23, 1986, is the main regulation on EIAs. It sets criteria for EIA requirements, defines

Table 2.1 Environmental Governance in Brazil

*National Environmental System (*Sistema Nacional do Meio Ambiente—*SISNAMA)*	
National Environmental Council (*Conselho Nacional de Meio Ambiente*—CONAMA)	It is an advisory and deliberative body. It proposes environmental guidelines and standards at the national level.
The Brazilian Institute of Environment and Renewable Natural Resources (*Instituto Brasileiro do Meio Ambiente e dos Recursos Naturais Renováveis*—IBAMA)	It implements CONAMA's resolutions. It reviews EIAs and issues environmental permits at the federal level.
State and local environmental agencies	They implement CONAMA'S resolutions. They review EIAs and issue environmental permits at state and local levels.

Source: Clovis Cavalcanti, "Economic Growth and Environmental Protection in Brazil: An Unfavorable Trade-off." *Decentralization, edited by Albert Breton Et. All, Edward Elgar* (2007); And Law N. 6938 of 1981 (Brazil).

environmental impact, basic activities and content of EIA, and its environmental impact statement (EIS), which is the report that results from the EIA procedure.[21] Milaré points out a contradiction in CONAMA Resolution No. 001/86. Although the resolution says it regulates Environmental Impact Evaluation, it only deals with EIA.[22] The Environmental Impact Evaluation is the broad term which encompasses different environmental studies, which range from simpler studies to the full EIA.[23] This limited scope of CONAMA Resolution No. 001/86 (restricted to EIAs) has led practitioners and experts to conclude that the Environmental Impact Evaluation refers only to EIA.[24] Milaré points that this is probably one reason why governments took a long time to require different methods of Environmental Impact Evaluation.[25]

The consequence of this wrong interpretation was the fact that agencies required EIA, which is a complex, expensive, and long procedure, in cases where a simpler environmental study could have been enough. Thus, it implied that more time and money spent in situations that the full EIA was not necessary. In this regard, Beltrão states that the EIA cannot be a "panacea" for the assessment of any environmental impact.[26] Furthermore, Article 2 of CONAMA Resolution No. 001/86 states that activities that modify the environment will be subject to EIA and environmental permitting procedures. According to Beltrão, however, Article 2 is vague and contains a technical error because even the simple act of cutting a tree modifies the environment.[27] However, the act of cutting one tree is not subject to EIA.[28] Despite these facts, the CONAMA Resolution No. 001/86 is, up to date, the main federal regulation on EIA.

CONAMA Resolution No. 006, of September 16, 1987, regulates major activities, especially those which are of interest of the federal government, such as power generation.[29] In general terms, this resolution sets criteria for power plants that were already operating or engaged in environmental permitting procedure when the National Environmental Policy was enacted and required Environmental Impact Evaluation for activities that would possibly cause significant harm to the environment.[30]

CONAMA Resolution No. 009, of December 3, 1987, regulates public hearings, which is the opportunity to inform the public on the findings of the EIA. Public hearing is not obligatory; however, a hearing is compulsory if requested by fifty or more citizens, the public ministry, or an institution of civil society.[31] Otherwise, the environmental agency invalidates the entire environmental permitting process and consequently the permit, if it was issued without holding the public hearing previously required by the mentioned actors.[32]

After a long period of military dictatorship, in 1988, Brazil promulgated a new democratic constitution. The 1988 Federal Constitution is also known as "the green constitution" because it established one chapter dedicated to the environment.[33] It was the first Brazilian Constitution to guarantee a right to an ecologically and balanced environment.[34] It also recognized the environment as a common asset, which the government and community have the duty to defend and preserve for present and future generations. Thus, an ecologically balanced environment constitutes an essential fact to the human right to life.[35]

The environment is neither public nor private. It is a diffuse right. Therefore, it does not belong to the union, states, or municipalities. However, the responsibility to manage and protect the environment for the benefit of all rests on all entities of the federation.[36] Therefore, the Federal Constitution establishes measures to guarantee the right to an ecologically balanced environment, such as the environmental impact study. The constitution refers to environmental impact study having the same meaning of environmental impact evaluation that was used in CONAMA Resolution No. 001/86. Article 225 states that

> All have the right to an ecologically balanced environment, which is an asset of common use and essential to a healthy quality of life, and both the Government and the community shall have the duty to defend and preserve it for present and future generations. Paragraph 1. In order to ensure the effectiveness of this right, it is incumbent upon the Government to: IV—demand, in the manner prescribed by law, for the installation of works and activities which may potentially cause significant environmental degradation, a prior environmental impact study, which shall be made public.[37]

Therefore, the requirement of carrying out an EIA, which is one modality of environmental impact study, is a constitutional mandate. It is obligatory whenever an activity may potentially cause significant degradation of the environment and the constitution carefully expressed that the study should be carried out prior to any action. Additionally, it determines that such documents shall be made public. The constitution fixed the technical error in CONAMA Resolution No. 001/86 by determining that environmental impact study will be required whenever the activity may potentially cause significant degradation of the environment. This is in contrast to the language of the referred CONAMA Resolution, which merely required that the activity would be able to modify the environment.[38]

On December 19, 1997, CONAMA issued Resolution No. 237, of December 19, 1997. It regulates environmental permitting procedure, such as competence of federal, state, or local agencies to conduct and consequently review, EIAs. CONAMA Resolution No. 237/97 defines environmental studies as all studies concerning the effects caused by the localization, installation, operation, and expansion of an activity or enterprise. Such studies serve as a basis to analyze the environmental permit required. The resolution provides some examples of environmental studies, such as environmental reports, management plan of environmental control, preliminary environmental report, environmental diagnosis, recuperation management plan of degraded areas, and preliminary risk analysis.[39] Having defined environmental impact studies, CONAMA Resolution No. 237/97 clarified the narrow scope of CONAMA Resolution No. 001/86, which suggested the erroneous idea that environmental impact evaluation is the same as EIA.[40]

The provisions of CONAMA Resolution No. 237/97 on competence to conduct the environmental permitting procedure had raised, for many years, questions and doubts on what agency would be responsible for issuing the environmental permit (federal, state, or local agencies) in each specific case. For instance, Article 4, III, sets forth that IBAMA (the federal agency) would be competent for conducting environmental permitting procedure of projects or activities when impacts would affect more than one state. The same criteria were applied to states, which would be responsible for conducting environmental permitting procedures when the environmental impacts would affect more than one municipality. The referred resolution adopted the extension of damage criteria.[41] These provisions did not shed the light on which would be the right agency to review the EIA and issue the permit, because commonly, impacts are not limited to only one state or only one municipality.

Additionally, scholars have pointed that CONAMA Resolution No. 237/97 was unconstitutional because only a law (and not a resolution) could regulate competence to conduct environmental permitting procedure, according to Article 23 of the Federal Constitution, which establishes that a Federal Law would define environmental competences of the union, states, and municipalities.[42] However, it was the only norm establishing competence in the environmental arena. Therefore, until 2011, practitioners applied this resolution to define competences of federal entities to issue environmental permits. Entrepreneurs commonly submitted the proposal of project and requested the environmental permit at the federal, state, and local agencies simultaneously. Entrepreneurs did so to avoid legal actions against them regarding the competence of the agency to conduct the environmental permitting procedure.[43] This lack of clarity related to environmental competence caused significant hardship to companies.

The Rio de Janeiro Petrochemical Complex (in Portuguese *Complexo Petroquimico do Rio de Janeiro* [COMPERJ]) clearly exemplifies this lack of clarity about environmental competence. COMPERJ is a major oil refinery in Rio de Janeiro, which had its environmental permits canceled and civil construction stopped while in the advanced stages. COMPERJ was paralyzed because the public ministry filed a legal action alleging the incompetence of the state environmental agency to conduct the environmental permitting procedure of the project. This resulted in financial loss for Petrobras—the company responsible for the project.[44]

Aiming to resolve the long-standing problem of environmental competence, in 2011, Congress enacted Federal Law No. 140, of December 14, 2011, which regulates environmental competence among different levels of the federation as stated in the Federal Constitution (Article 23).[45] The constitution states that all entities of the federation (union, states, and municipalities) have shared competence to protect the environment as well as preserve the fauna and flora.[46] Thus, no hierarchy exists among them for environmental protections.[47] In this regard, according to Article 23, union, states, and municipalities have competence to conduct environmental permitting procedure and review EIAs.

Law No. 140/11 clarified the competence of IBAMA (the federal agency) to conduct environmental permitting procedure. Article 7 (XIV) notes that IBAMA is competent to conduct environmental permitting procedure, and consequently review EIAs, when the activity or project is located (1) in Brazil and in a neighboring country; (2) in the territorial sea, continental shelf, or exclusive economic zone; (3) on indigenous lands; (4) in protected areas established by the federal government; (5) in two or more states, among others. It takes into consideration the localization of the project and not the extension of the environmental degradation, as it was adopted by CONAMA Resolution No. 237/97.

Concerning states, they should conduct environmental permitting procedure whenever the competence is not allocated to the union or municipalities. Finally, municipalities will conduct such procedure whenever the project or activity will have a local impact. Regarding municipalities, the referred law kept the environmental degradation criteria, noting that state-level environmental council will decide on the definition of local impacts without the input or participation of municipalities.[48] Thus, Lima points out that the Federal Law is unconstitutional because it subsumes the competence of the municipality to a decision of the state, which breaches the constitutional mandate of cooperation among all entities of the federation.[49] Considering that there is not participation of the municipalities in the decision of local impacts, there is an arbitrary provision in Law No. 140/11.

Other scholars, such as Souza, state that Federal Law No. 140/11 incentivizes economic and unsustainable growth because when it defines that municipalities will be responsible for conducting environmental permitting procedures, it opens the door to corruption and approval of poor EIAs. This is because some municipalities do not have personnel qualified to analyze EIAs.[50] Some municipalities are extremely poor. Thus, they might not even have agencies exclusively dedicated to the environment. For instance, Belágua, which is a municipality of the State of Maranhão, in the northern region of Brazil, is considered one of the poorest Brazilian municipalities. It has 569,428 square kilometers of extension, with a population of 6,524 people, where 52.11 percent are illiterate, and, the monthly income ranges from R$30.00 to R$699.56, which approximately corresponds to US$10 and US$225.82 respectively.[51]

Guerra points out that, in practice, municipalities will barely conduct environmental permitting procedures. Considering that it is under the state's council discretion to decide on local impacts, states will normally decide that they have the competence to conduct the permitting procedures.[52] Glasson states that whether municipalities have the technical capacity to conduct environmental permitting procedures, local stakeholders would contribute to the debate and decision-making of the project. However, Glasson and Salvador acknowledge the lack of capacity of some Brazilian municipalities.[53]

Therefore, on the one hand, Federal Law No. 140/11 established clearer baselines on the competence of states and union to carry out environmental permits than the previous CONAMA Resolution No. 237/97. On the other hand, it guarantees competence to municipalities which does not work in practice for the following reasons: (1) many Brazilian municipalities do not have personnel to review EIAs and (2) concept of local impact is relative and

Table 2.2 EIA Legislation in Brazil

	Legislation Related to EIA in Brazil
The 1988 Federal Constitution	It states that Environmental Impact Studies will be required whenever the activity may potentially cause significant environmental degradation. The environmental impact study shall be made public (Article 225, IV).
Law 6938/81	It establishes the environmental governance and states that the permitting procedure and the EIA are tools of the National Environmental Policy.
CONAMA Resolution N.001/86	It provides General Guidelines for the Implementation of Environmental Impact Evaluation.
CONAMA Resolution N.006/87	It sets criteria for power plants which were already operating or engaged in environmental permitting procedure when the National Environmental Policy was enacted and required Environmental Impact Evaluation for activities which would possibly cause significant harm to the environment.
CONAMA Resolution N.009/87	It regulates public hearings within environmental permitting procedure.
CONAMA Resolution N.237/97	It regulates environmental permitting procedure as established in the National Environmental Policy Act.
Federal Law N. 140/11	It deals with environmental competence among different levels of the federation to conduct permitting procedure and review EIAs.

considering that it is under states' responsibility to define whether the impact is local or not, states will probably define that the impact is not local and consequently the state will conduct the environmental permitting procedure.

Overall, Federal Law No. 140/11 exemplifies the lack of legal clearness on a very important issue, which is the competence to review EIAs. This is a major problem that affects the development of EIAs in Brazil because it results in legal actions, delaying the normal timeline of EIA procedure and of the project. These unexpected delays might prejudice the development of EIAs and result in political pressure to approve the project independent of adverse effects or the quality of its EIS. The case study, which is examined in chapter 3 is one example of conflict of competence between federal and state environmental agencies.

BRAZILIAN ENVIRONMENTAL PERMITTING PROCEDURE AND EIA

EIA approaches vary among countries. Some require EIAs in mandatory regulations, acts, or statutes. In these cases, the realization of EIA is a precondition to approval of projects, policies, or programs, and, normally, EIA happens in the scope of land planning and/or environmental permitting, such as in Europe. Conversely, some countries have EIA guidelines which are not obligatory. Other legislations establish that it is under governmental authorities' discretion to require EIA before approval of development. In other cases, EIA is an ad hoc practice, required by financing agencies such as the World Bank.[54] Glasson states that it is not the EIA system that determines its effectiveness. For instance, Brazil and the Philippines are recognized as countries where the EIA regulations are not well-enforced in practice.[55] Whereas EIA guidelines work well in Japan and, in the UK, some ad hoc EIAs have been well-prepared.[56]

The main strength of having EIA attached to the planning or permitting process is the fact that it is more difficult to avoid its realization.[57] In Brazil, the EIA happens within the Environmental Permitting Procedure, according to Article 1 of CONAMA Resolution No. 237/97.[58] It is a condition to the issuance of the environmental permit required. The National Environmental Policy Act (Law No. 6938/81) requires that "The construction, installation, expansion and the operation of activities which will utilize natural resources, considered effective or potentially polluting as well as capable of causing environmental degradation must be preceded by Environmental Permitting Procedure."[59] Decree No. 99.274/90 regulates this requirement and establishes the world's unique system of three-stage permitting process,[60] which encompasses the following environmental permits: preconstruction, installation, and operation.

The preconstruction permit approves the design and the localization of the project in accordance with the governmental land use planning; it also authorizes the environmental feasibility of the project. Additionally, it establishes conditions to be fulfilled by the project proponent in the subsequent phases of the process.[61] The installation permit authorizes the beginning of civil constructions according to the plans, programs, or projects approved, including the environmental controlling measures.[62] Finally, after the environmental agency verifies that all the conditions established in the previous permits were fulfilled, including the environmental controlling measures, it issues the operation permit, which authorizes the beginning of the activities or enterprise.[63]

Thus, the environmental permitting procedure is only one procedure; however, it is divided into different stages resulting in three permits, which authorize distinct phases of project implementation. Compliance with one stage is a condition of the next one. In other words, noncompliance with conditions established in one permit implies the environmental agency's rejection to issue the next permit. In daily practice, even if conditionalities are not fulfilled in one permit, the environmental agencies often issue the next permit restating the same conditionalities.[64] This practice breaches the provisions contained in CONAMA Resolution No. 237/97 and weakens the entire system because it postpones obligations without complying with them in appropriate time.[65]

Annex I of CONAMA Resolution No. 237/97 lists the projects that must go through an environmental permitting procedure. For instance, open-pit mining, steel mill, chemical productions, among others. It is true to affirm that EIA always occurs within the Environmental Permitting Procedure; however, the Environmental Permitting Procedure will not always require an EIA.[66] The competent environmental agency will evaluate the necessity of requiring EIA or any other environmental study necessary to issue the environmental permit. Article 3 of the referred resolution states that

> The environmental permit for enterprises and activities, which will effectively or potentially cause significant environmental degradation, will depend on prior environmental impact assessment and its respective environmental summary report. . . . If the competent environmental agency finds that the activity or enterprise will not potentially cause significant environmental degradation; it will define the environmental studies pertinent to the environmental permitting.[67]

The following question arises: Considering these three stages of permitting procedure, in which phase does the EIA occur? The EIA must be developed before the preconstruction permit because it provides information to analyze the feasibility of the project in the specific location required by the project proponent and consequently grounds the decision of the permit granted.[68] In

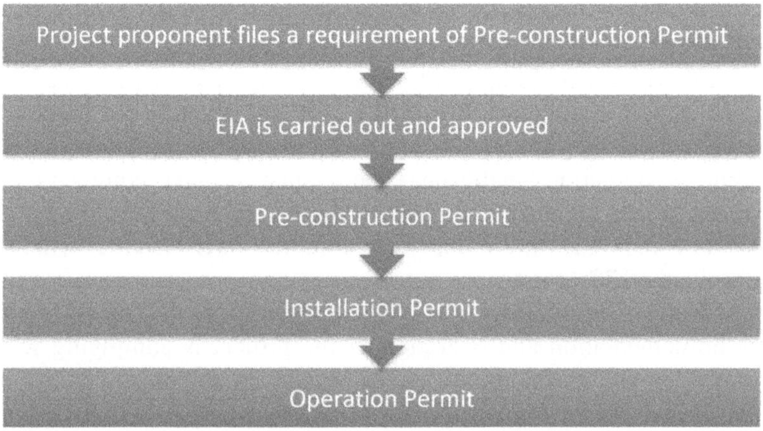

Figure 2.1 The EIA within the Environmental Permitting Procedure.

this regard Machado points that the Federal Constitution determines that the environmental impact study must be developed prior to the project's authorization.[69] Therefore, EIA cannot be carried out concomitant with or after the installation and operation of projects.

The flow chart in figure 2.1 demonstrates the moment in which EIA happens during the environmental permitting procedure.

These are the legal steps required at the national level for constructing a project which might cause significant environmental impact. However, at the state level, some states have their own EIA regulations, which might have other requirements depending on the project.

EIA PROCEDURE IN BRAZIL

Several steps compound EIA procedures around the world. There is not one EIA system standard that is followed by all countries. Conversely, it depends on each country's policy or EIA approaches on what steps the EIA will cover. Furthermore, the literature descriptions of EIA steps vary. For instance, Glasson provides a detailed classification of EIA procedure,[70] while Wood recognizes a shorter procedure.[71] This book adopts the following EIA stages. These are the main phases of the procedure in most countries, including Brazil:

1) Plan the project (prepare a proposal);
2) Screening—decision whether EIA is necessary for that specific project or activity;

3) Scoping—identification of all potential impacts of the project the EIA will cover;
4) Write Environmental Impact Statement (EIS)—document all the findings of the impact assessment. For instance, it will describe the proposal, the baseline area, the impacts of the project and propose mitigation measures;
5) Public participation—public has the chance to make comments (either written or oral comments in public hearings) on the EIS findings;
6) Decision-making—the competent authority will decide whether to approve the EIS or not;
7) Post-decision monitoring—the competent authority will monitor whether the implementation and operation of the project is complying with the controlling and mitigation measures previously determined.

There are variations depending on the country. Additionally, not all phases happen in all EIA systems.[72] Although the procedure is described as a list, it is cyclical and not linear.[73] For instance, after public participation and consultation, the project proponent might need to redesign the project proposal.[74] In the United States, the Council on Environmental Quality (CEQ) Regulations define each of these phases.[75] However, in developing countries, normally the EIA procedure is part of a broader environmental law and it does not have a detailed process requirement.[76]

In Brazil, CONAMA Resolution No. 237/97, in broad terms, describes the environmental permitting and touches on the EIA procedure. CONAMA Resolution No. 001/86 states which projects require EIA, the content of the EIS, its respective environmental impact summary report, and the necessity of making these documents available to the public. Thus, the EIA procedure phases are not formally defined in Brazilian legislation. It is vaguely mentioned in the scope of the environmental permitting procedure, since it is part of it.

The chart in figure 2.2 briefly describes the Brazilian EIA procedure.

The project proponent develops the proposal of a project in a specific location. The proponent requires from the competent environmental agency (state, federal, or local agency) the issuance of the preconstruction permit. The environmental agency evaluates the necessity of carrying out EIA or any other environmental study, based on the illustrative list of projects in CONAMA Resolution No. 001/86 (screening phase). The environmental agency in conjunction with the project proponent will draft a Terms of Reference (TOR), which will outline the points that the EIA will cover.

The project proponent will hire a team of consultants who will collect data and eventually write the EIS followed by an environmental impact summary report, which should be written in nontechnical language to inform the public about the proposed project. After receiving the EIS, the environmental agency

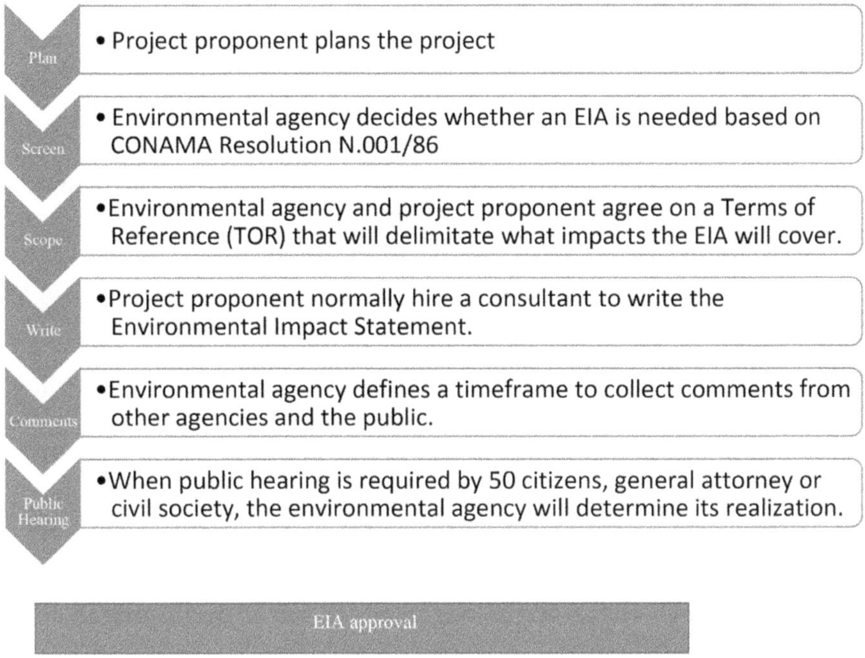

Figure 2.2 EIA Procedure in Brazil.

will establish a written comments period, which will receive comments from other agencies as well as from the public.[77] The environmental agency has discretion to decide on carrying out a public hearing or not. However, whether *Ministério Público*, a group of fifty or more citizens, or a civil society organization requires it, the realization of public hearing is obligatory.[78]

The environmental agency might request further clarification as a result of the public hearing. It will issue a technical and sometimes a legal opinion which will approve or deny the project.[79] Whether the project is approved, the preconstruction permit will be issued containing conditions that must be fulfilled by the project proponent in order to be granted the other environmental permits. Before the issuance of the installation permit, project proponent must present the Basic Environmental Plan (in Portuguese *Plano Básico Ambiental* or PBA), which details mitigation and compensation measures in relation to impacts identified in the EIA.[80] Only after project proponent proves that it has complied with conditionalities stated in the preconstruction permit and it files the referred plan, the installation permit is issued.[81] The environmental agency will monitor compliance with the conditions stated in the permits, which reflect the findings of the EIA.

As per responsibilities, project proponent can carry out EIA.[82] However, normally, companies do not have necessary expertise within their staff. Therefore, they directly hire consultants to develop the EIA. For instance, an EIA of a mining project requires at least a geologist, a biologist, an archaeologist, depending on the area, an anthropologist, and/or an expert in social sciences.[83] Consultants and project proponents have regular meetings during the realization of the EIA to share information on the project and its impacts, analysis and field visits. Both project proponents and consultants might meet with environmental agency as well to present the stage of the EIA during its construction. However, these are informal meetings, not prescribed in legislation.

EIA Timeline

At the federal level, IBAMA Executive Order (in Portuguese *Instrução Normativa*) No. 184, of July 17, 2008, amended by Executive Order No. 14, of October 27, 2011, determines the following deadlines:[84]

- TOR—IBAMA has sixty days to issue the TOR for EIA. The TOR is valid for 2 years
- Checklist—IBAMA has thirty days to check adequacy of the EIS in relation to the TOR
- Review—IBAMA has 180 days to review the EIS
- Publish the proposed EIA in the newspaper and official gazette—public hearing must happen after forty-five days

These deadlines refer to EIA being carried out by the federal agency—IBAMA. At state or municipal levels, environmental agencies might have their own regulations; if not, they can apply the deadlines mentioned in the list.

Brazilian EIA deadlines seem to be very ambitious, at least on paper. For example, in Canada the review of EIAs carried out by the Impact Assessment Agency is 300 days.[85] In comparison to the 180 days stated in Executive Order No. 184/08, Brazilian regulation allows a very short period of time to review EIAs. However, in practice, IBAMA does not comply with deadlines stated in Executive Order No. 184/08. This happens for several reasons, such as lack of qualified personnel and frequent court interventions determining that environmental permitting procedure should be suspended. EIA timelines are discussed further in this chapter.

Environmental Permitting Procedure and EIA of Mining Activities

The EIA procedure is the same for all activities in Brazil; however, in relation to certain activities, environmental permitting can happen after or concomitant with other procedures involving different governmental agencies. For instance, oil and gas activities require previous intervention of the National Agency of Petroleum; mining activities require previous approval of the National Department of Mineral Production (in Portuguese *Departamento Nacional de Produção Mineral* [DNPM]). The main problem in these cases is poor communication between these sectoral and environmental agencies.

Mining is the activity of extraction and processing of mineral substances from deposits. Mining projects have four different phases that involve DNPM and the environmental agency: research, installation, operation, and decommissioning or closure of the mine.

Before engaging in a research phase or mineral prospection, project proponents must seek authorization from DNPM. The goal of mineral prospection is to ensure the existence of minerals in the deposits, previously identified by secondary sources. Initially, prospection involves methods that do not result in environmental impact or result in minimum impacts, such as flying the area of the deposit to collect necessary information and site visits carried out by geologists. Normally, these initial steps are followed by soil perforation and sometimes the collection of a certain amount of minerals for further studies. In this case, DNPM must issue an authorization specifying the quantity of minerals to be extracted.[86]

In addition, project proponents will also need to engage in environmental permitting procedures to perforate and/or collect minerals from the deposit as these actions will probably cause social and environmental impacts. Therefore, the environmental agency issues an operation permit allowing interventions in the area during the research phase. At the end of the mineral prospection phase, project proponents present a final report to DNPM. The next step in this process is the submission of a mineral financial plan to DNPM. If this plan is approved, the project proponents carry out the EIA and write the EIS under the auspices of the competent environmental agency. The EIA will follow the normal path (scoping—draft the TOR, public hearing if required, decision).[87]

Once the environmental agency approves the EIA and the project, it issues the preconstruction permit. Project proponents must comply with the conditionalities stated in the preconstruction permit and submit an environmental monitoring plan to the environmental agency, which will approve or reject the plan. These are conditions for the issuance of the installation permit.[88]

Only after the issuance of the installation permit can the project developer request a mining concession to DNPM. Project proponents are allowed to start construction of the project, according to the environmental monitoring plan presented and conditionalities stated on the installation permit. Both DNPM and the environmental agency will monitor the installation of the project. Then, after completion of civil constructions, the environmental agency will analyze whether all conditionalities were fulfilled and, furthermore, whether the installation complied with the environmental monitoring plan before issuing the operation permit.[89] Project proponents can only start operating the mine when the operation permit is issued.

As described here, there is an intersection of mining concession and environmental permitting. Therefore, an effective communication among these agencies, project proponents, and consultants is important to guarantee effectiveness of these procedures. However, in practice, there is often no coordination between actions involving both processes. This fact results in pressure to rush the environmental permitting because project proponent has already spent some time securing authorization for the prospection phase.[90]

IBAMA's guidelines on environmental permitting procedures of mining activities points out that the project proponent carries out EIA at a late stage, in relation to the concession procedure. This means that the project proponent carries out EIA and requests the preconstruction permit after the mineral financial plan is approved by DNPM. Instead the EIA should have started during the prospection phase; therefore, both procedures (mining concession and environmental permitting) would be conducted in parallel, which would save time and assess impact more effectively. Furthermore, due to this lack of coordination, project proponents tend to complain that the environmental permitting procedure is long and bureaucratic.[91]

Although this is an important issue, this book does not analyze mining concessions and environmental permitting procedures further. The goal is to demonstrate that there are actions and procedures conducted by DNPM prior to the environmental permitting procedure of mining activities. There are coordination problems that can result in complaints, perception of unnecessary bureaucracy, and long periods of time to implement projects. Further studies are necessary to enhance the coordination of procedures and optimization of time. It is also very important to emphasize that although there is interaction with DNPM, the EIA of mining activity is exactly the same procedure for any other activity. EIAs for mining activities tend to follow the same stages as EIAs for oil and gas production, construction of roads, or installation of a cement factory for example. In sum, the stages of screening, scoping, public participation, and decision are the same for all activities. An example presented in chapter 3—a mining activity—provides a brief explanation of the role of DNPM and the environmental permitting procedure to seek authorization for mining projects.

Public Participation in Brazilian EIAs

An extensive body of literature as well as practitioners point out the benefits of having public input on the assessment of potential impacts of programs, policies, and projects.[92] From the point of view of governments and proponents of projects, public participation can be a positive step toward the achievement of their goal: implementation of projects. Copious literature illustrates this point: local communities may provide valuable and realistic information concerning the area where a project is planned to be implemented, which assists the development of an EIA and therefore the project's financial plans. Moreover, it increases people's credibility, which diminishes chances of haunting the project by protests, strikes, and legal actions.

Considering the people's side, public participation in governmental decision-making is internationally recognized by several soft laws and conventions, such as the 1992 Rio Declaration on Environment and Development and the American Convention on Human Rights.[93] It is also a matter of social justice and environmental democracy to hear those who will face environmental impacts daily, whether from pollution, increase of traffic and population, water scarcity, loss of housing and social ties, among other effects of development projects. Therefore, EIA should truly consider not only experts' opinions but also all stakeholders' views and concerns to achieve sustainable decisions on approving or rejecting a project.[94]

A study from the Harvard Kennedy School concluded that social investment and the quality of consultation processes decrease conflicts between communities and project developers. The research was based on an analysis of fifty worldwide extractive industry projects and demonstrated that the most frequent and major costs faced by extractive industries are those related to temporary shutdowns or delays which impact productivity. "For example, a major, world-class mining project with capital expenditure of between US$ 3-5 billion will suffer costs of roughly US$ 20 million per week of delayed production in terms of Net Present Value (NPV), largely due to lost sales."[95] In this context meaningful public consultation plays a crucial role in reducing the risks of financial losses.[96]

Ganson states that "a study of direct costs to business of company community conflicts found one nine-month construction delay that resulted in US$ 750m in additional project costs. Stoppage at another company's project carried a price tag of US$ 100m per year, and a community's ability to protest against a third company by cutting its power lines cost it US$ 750,000 per day."[97] In Peru, Newmont mining reported that it experienced loss of approximately US$2 million per day, after protests paralyzed its Conga mining project.[98]

Furthermore, a 2011 empirical study of nineteen mining companies (between 1993 and 2008) on the financial gain of community engagement in relation to gold mining projects concluded that, "[i]n short, the social license to operate is more than rhetoric. It is operationalizable, empirically testable and strategically relevant. For these mining firms and their peers in politically and socially salient sectors, pursuing support from external stakeholders is not just corporate social responsibility but enlightened self-interest."[99]

Therefore, data demonstrates, in financial terms, the impact of, lack of, or weak public participation in decision of authorization of projects. In this context, the IFC 2012 Sustainability Policy states that increasing the overall sustainability of operations their clients would build competitive advantage in the marketplace.[100] Therefore, the cost-analysis benefits of projects should include participation in decision-making which in turn results in financial gain.

In Brazil, counterarguments to extensive public participation in EIAs are time and cost. The environmental permitting procedure, which encompasses the EIA, is already long. The private sector states that there is too much bureaucracy involving the referred procedures and it can take up to five years to obtain an environmental permit, which includes conducting an EIA.[101] It is a common belief that due to excess of bureaucracy and considering the normal timeline to obtain a project's approval, national and international enterprises feel discouraged to invest in the country.[102] In relation to cost, private enterprises state that there are already costs associated with the development of the EIA, realization of public hearing, mitigation, and compensation measures.

Are Time and Cost of EIA Barriers to Enhancing Public Participation in Brazil?

The impact on cost depends on the complexity of the project and its location. The WB and the IDB indicate that the costs of environmental assessments rarely overcome 1 percent of the total cost of the project.[103] The WB points out that the cost of an environmental impact analysis, in general, ranges from 0.06 percent to 0.10 percent of the total cost of the project, considering small to large-scale projects respectively.[104]

In relation to timeline, a WB report states that EIA procedures around the world might take from one to three years. However, in Canada, the permitting procedure for large-scale hydroelectric dam can last five years. The permitting procedure is understood here as the time of filing of the TOR of the EIA until the final approval of the project. The EIA itself, including all its phases, takes three years. While a medium-size hydroelectric dam permitting procedure might take three years in total, two years may only be dedicated to the EIA.[105]

In relation to the usual complaint that EIAs are time consuming, Garry and Isaac Middle state that, considering the lack of data on the timeline of EIAs, practitioners base such a statement more on perception than on facts. They analyzed full EIA procedures from 2000 to 2010 in Western Australia, demonstrating that, in general, full EIAs took an average of 567 days in Western Australia.[106]

Furthermore, research carried out by University of Oxford professors analyzed 245 large dams around the world between 1934 and 2007, including the Belo Monte and Itumbiara dams in Brazil. The professors concluded that project proponents do not plan time and cost accurately, especially when they seek governmental authorization. Costs and time are underestimated because real data would risk questioning the economic feasibility of these large projects. In relation to cost, the study found that dam price forecasts should be increased by 57 percent just to provide an 80 percent certainty that the final cost of the project would be within the initial budget. They also found that "actual costs were an average 96% higher than estimated costs, the median was 27%. The evidence is overwhelming that costs are systematically biased toward underestimation."[107] Costs of implementation of these projects also increase due to delays in the schedule (also often underestimated) and unpredictable and unquantifiable events, such as stoppages and protests.

In Brazil, the environmental permitting procedure of a large-scale hydroelectric power plant normally takes six and a half years, with high variances, according to the WB report.[108] This timeline is 30 percent longer than in the United States and one and a half years longer than in Canada.[109] Considering that Brazil is a developing country—presenting several challenges, including lack of personnel in environmental agencies able to review EIAs, and the existence of many complex social and environmental issues in areas where projects are implemented—the difference in time is not significant.

The World Bank's report, "Environmental Licensing for Hydroelectric Projects in Brazil," concluded that the three stages of the environmental permitting procedure "generate much uncertainty, *lengthy delays* and high transaction costs" (emphasis added).[110] Conversely, research carried out by the Study Group on Environmental Issues of the Federal University of Minas Gerais revealed that the environmental permitting procedure should not be more flexible.[111] Instead, the practice appears problematic, where the environmental permit is issued with conditionalities to fix problems that should have been resolved before the issuance of a permit. Additionally, the main reason for delays is the poor quality of EIAs, which are incomplete and conducted without transparency.[112] This fact requires clarification from both the environmental agency and *Ministério Público* by filing legal actions.[113] Although there is no consensus on which system would be more appropriate (three- or one-stage environmental permitting procedure) to protect the environment, in Brazil, the conditionalities stated in one permit are, not rarely, repeated in the next permit without being fulfilled. One example is the Hydroelectric Dam Jirau, in Rondonia, in the northern region of Brazil. The installation permit was issued for Jirau even though only twelve out of thirty-two conditionalities stated in the previous permit were fulfilled.[114] Therefore, such practice weakens the purpose of having a three-stage environmental permitting procedure.

A 2015 study prepared by Hofmann for the Brazilian House of Deputies points out that frequent postponement of a permit's conditionalities is one of the causes of judicialization of EIAs.[115] Other reasons for a great number of legal actions involving EIAs are discussions on the competence of the state or federal environmental agency to review EIAs and the lack or insufficient public participation.[116]

In Brazil the *Ministério Público* has a distinctive role.[117] It is compounded by public prosecutors and is independent from any other power (judiciary, legislative, or executive). The Brazilian Federal Constitution states that the mandate of the *Ministério Público* is to defend the juridical order, the democratic regime, and the social and inalienable individual rights.[118] It is responsible for filing legal actions to defend environmental interests and other collective and diffuse rights.[119] Although other actors are entitled to file legal actions in protection of the environment,[120] in most of the cases, the *Ministério Público* does so.[121]

Considering the great number of interventions of the *Ministério Público* in environmental permitting procedures, including EIAs, there are critiques stating that it has excessive power, and its constant intervention delays development. For instance, the WB report states that "The [*Ministério Público*] has not employed its resources to solve problems but has instead come to represent an additional and controversial impediment to the environmental licensing of major developments, especially hydropower plants."[122] Contrarily, Hofmann states that the environmental permitting procedure has several weaknesses, such as institutional and legislative uncertainties as well as lack of transparency.[123] Therefore, it is not right to affirm that there is excessive intervention of the *Ministério Público* in the Environmental Permitting Procedure.[124] In fact, intervention of the *Ministério Público* is necessary in order to guarantee the preventive nature of EIAs, by arguing in court every time the procedure does not comply with the law—which occurs often. Furthermore, social conflicts are rarely resolved during the environmental permitting procedure, which leads to further delays in the approval of projects in Brazil.[125] Therefore, the critique that EIAs are costly and time consuming does not stand.

In summary, costs of EIAs do not cause significant financial impact on projects. EIA procedures are lengthy in Brazil due to shortcomings in implementation, including poor quality EISs and lack or insufficient public participation, which results in legal actions filled by the *Ministério Público.* However, an extensive process that strengthens public participation throughout the EIA would minimize conflicts and enhance social control over the quality of EISs. Consequently, lengthy delays could be avoided. Robust public engagement in EIAs reflect a good image of projects that might impact the sustainability indicators and prices of shares. In addition, it diminishes protests that tend to haunt the project, along with legal actions questioning several aspects of the projects and the EIA, as previously mentioned.

The next section briefly describes how EIAs are regulated in different countries, specifically their legal requirements on public participation.

COMPARATIVE ANALYSIS OF EIA IN DIFFERENT COUNTRIES

EIA is a planning tool that, if well utilized, can be effective in deciding the social, environmental, and economic future of regions and nations. However, it has shortcomings in different parts of the world.[126] While public participation is not a problem in the United States,[127] criticisms highlight the fact that agencies responsible for the authorization of projects are not bound by public input nor by the findings of EISs. It is merely a procedural obligation.[128]

In Finland, the EIA legislation states that project proponents are not obligated to minimize negative environmental impacts resulted from projects. Polonen also points out that alternatives to the project are not truly considered in Finnish EISs. On the positive side, public participation increased in 1997 and 2003 with amendments to the EU EIA Directive.

In developing countries, legislation might contain the very basic steps for public participation in EIA's legislation; however, implementation is still deficient. Argentina has a national environmental policy, Law No. 25.675/2002. It establishes the basic standards regarding environmental matters, including the realization of EIAs, which are carried out by the provinces.[129] Nearly all provinces have EIA regulations and public hearings are obligatory for projects that might significantly affect the environment.[130] However, Verocai states that implementation of the EIA procedure in Argentinian provinces is very slow due to the lack of human resources and the unstable political and institutional situation of the Environmental Council.[131]

In Bolivia, projects listed on the positive list will be subject to EIA. Public hearings are obligatory in these cases. If a hearing is not carried out, the environmental authority shall subject the draft EIS to public consultation, when written comments will be collected and considered before the issuance of the EIS.[132]

In Chile, Law 19.300 of March 9, 1994, and Decree No. 40/2013, which came into force in October 2014, regulate EIA practices.[133] The Decree requires that the project proponent publish the abstract of the EIS in the official gazette, in a newspaper of the capital city of the region, or of national circulation, and in the radio.[134] Public participation is generally limited to written remarks of any one person. While the written comments are reasonably well-described in the legislation it is not explicit about public hearings or meetings. Instead Law 19.300/94 states that the authority would decide about the mechanisms that will ensure public participation in the EIA procedure.[135]

In Uruguay, Decree No. 349/2005 determines that public participation is mandatory for certain projects or when it is required by a civil entity, the public prosecutor, or fifty citizens. Similar to the participation model adopted in Brazil, more than one public hearing may be held in Uruguay if it is deemed necessary. During the hearing, space is reserved for questions from the public.[136]

In Venezuela, public participation is not mandatory for projects subject to EIA. However, the Ministry of Environment and Renewable Natural Resources may request public consultation of the EIS. Written comments are accepted in this case with technical, scientific, and legal reasons which support them[137](table 2.3).

In terms of EIA regulations and procedures, especially considering public participation in EIAs, the process is fairly similar throughout South America. Brazil has decent written legislation regarding public participation in EIAs, in comparison to some countries—particularly in comparison to Venezuela, where comments should be legally justified. Or Chile, where the written comments period is reasonably well written; however, there is no explicit mention to realization of a public hearing or meeting.[138] In this vein, Rocha et al. state that among the Mercosul countries (Argentina, Brazil, Paraguay, and Uruguay), Brazil has more advanced environmental legislation.[139]

Additionally, a study carried out by the Environmental Democracy Index places Brazil sixteenth among seventy countries analyzed. The index measures legislation of countries in relation to access to information, and access to justice and public participation. Brazil performed well in access to

Table 2.3 Comparative Analyses—Public Participation in EIA

Country	*Legal Requirements/Notes*	*Public Participation*
USA	1969 NEPA §102	Public participation is well-informed
Finland	The 1997 and 2003 Amendments in the EU EIA Directive	Enhanced public participation in the country
Argentina	There is no national regulation	Public hearing is obligatory
Bolivia	Projects listed on the positive list will be subject to EIA	Public hearings are obligatory in these cases
Chile	Decree No. 40/2013	In general, written comments is the mean of public participation
Uruguay	Decree No. 349/2005 determines that public participation is mandatory for certain projects or when it is required	During the public hearing spaces are reserved for questions of participants
Venezuela	-	Public Participation is not mandatory

environmental information and access to justice, and fair or limited in terms of public participation. However, it should be noted that the index only evaluates national legislation on paper, thus, it does not consider practice.[140]

Although there is a general assertion that Brazilian environmental legislation is "strong," practice does not comply with this statement. While Brazil might have a fair environmental legislation in comparison to other developing countries—for example, in Egyptian EIA legislation, disclosure of the EIS and public participation is not mandatory[141]—Brazilian EIA legislation still needs to be enhanced and modernized, especially considering the disclosure of EIS and the momentum of public participation. Improvements in Brazilian EIA legislation would be an inspiration to other developing countries.

WHY IS PUBLIC PARTICIPATION IN EIAs WEAK IN BRAZIL?

In general, Brazilians do not actively engage in public participation processes as in full democracies.[142] Several reasons explain this fact, including historical and political circumstances, which are explained later in this section.

The Economist Intelligence Unit's Index of Democracy compares 167 countries worldwide in terms of their level of democracy. This index considers five categories: electoral process and pluralism, civil liberties, the functioning of the government, political culture, and political participation. Looking at the political participation, which includes voting in national elections, citizen engagement in politics as well as participation in political parties and nongovernmental organizations, Brazil scored 6.67 out of 10 in 2018, below the average of 7.83 from full democracies.[143] However, Brazil, considered as a flawed democracy, is not alone in Latin America. Even the two Latin American countries considered full democracies, Uruguay and Costa Rica, scored 6.11 and 6.67 respectively on political participation. In summary, on the topic of political participation, Brazil, Costa Rica, and Uruguay are on equal foot, according to the index.[144] Thus, political participation does not appear very strong in Latin America and Caribbean as a whole.

In relation to participation in formal processes to develop laws and regulations, the OECD survey found that the level of participation in this type of public engagement in Brazil is 2.2 out of 4, which is lower than the OECD average of 2.4.[145] Conversely, widely discussed topics seem to attract more participation in Brazil. In 2005, Brazil held a referendum consulting the population about the commercialization of firearms and ammunition to individuals. Among the 122,042,615 Brazilians who were apt to vote, 78.15 percent participated in this referendum[146]—a significant percentage of participation. When compared to the number of people who voted in the 2016 referendum in the UK, to determine whether the country should remain a member of the European Union, in which 72.2 percent of the population voted,[147] participation in Brazil is equally strong when the issue being discussed is well-known and of national interest.

Lack of knowledge of certain participatory mechanisms is also a problem that prevents the population's engagement. A study on participation in health committees in the south of Brazil demonstrates that, among the forty-six people interviewed in thirteen different municipalities, 69.2 percent did not know about the plans and objectives in relation to health issues in their city. This lack of knowledge reflected in low participation in health committees; only 13.1 percent participated in community meetings to discuss and plan actions related to health.[148]

Participation in EIAs seems to be weaker in some parts of Brazil than in others. For instance, comparing Rio de Janeiro—a big urban state—and Pará, which is located in the Amazon rainforest, local people participate more actively in the latter. Those who live in the area surrounding a project will be directly affected and, therefore, are more willing to provide input and to be heard by governments before a decision is taken. Normally, they are riverine peasants—poor and sometimes illiterate. Although they are willing to participate in EIA procedures, they often do not have access to EIS beforehand or the language used to explain the projects is too technical. For instance, it is estimated that 5,000 people attended the public hearing of Belo Monte held in Altamira, as opposed to the 300 people who attended a public hearing assessing the installation of two thermoelectric units in Macaé, Rio de Janeiro.[149]

Of course, the number of attendees at public hearings is not a good criterion to measure low or high participation because other facts influence participation in these events, such as advertisement of public hearings, geographic location, accessibility, and timing of hearings. Nor does it mean "meaningful participation" because hearings are not the most effective mean of participation, as explained throughout this book. However, participation rates demonstrate willingness (or lack of it) of Brazilians to engage in public processes that lead to important decisions.

Comprehensive study from the Brazilian Institute of Economic Research on the challenges of permitting procedure for urban development projects in Brazil was conducted between 2008 and 2010.[150] Twenty-three Brazilian municipalities' agencies provided reports and answered questionnaires, stating that public participation is low in permitting procedures of urban development projects, such as urban planning. This was the opinion of 60 percent of interviewees from permitting agencies, 77.8 percent of public entrepreneurs, and 69.2 percent of private entrepreneurs. In their opinion, as a consequence, the contribution of public participation in EIAs is also small. This was the opinion of 60 percent of permitting agencies, 72 percent of public entrepreneurs, and 70 percent of private entrepreneurs.[151]

Ultimately, low participation is not an exclusive problem to EIAs nor is it unique to Brazil. In Brazil, it depends on the profile of the project, the region, and the potential impacts on local population. Some reasons might explain weak public participation in Brazilian EIAs.

As a former colony of Portugal, and still a developing country, there is not a strong culture of participating in governmental decisions or encouragement of public debates.[152] Both project proponents and the government fear that public engagement will delay economic development; therefore, there is not much interest in strengthening transparency and public dialogue within EIAs. Furthermore, lack of trust in EIA outcomes and environmental decision-making as a reflection of distrust in governments also contribute to weak public participation. The following subsections provide some brief reflections on why public involvement and participation is poor in Brazil. These causes are not analyzed in depth; rather these reasons provide a clearer understanding of participation (and lack thereof) in Brazil.

Education

The lack of interest in participating in public decision-making can be explained by education methodologies.[153] The way classes are taught in Brazil, and in many other countries, does not actually encourage students to debate and participate in the classroom. Teachers explain the subject to the entire class (normally in forty-five minutes) and, in general, there is barely any discussion in classroom on the topic. Students can ask questions and provide their opinions; however, there is no incentive to a free debate. This fact demonstrates that the fundamental basis for participation does not rise from schools and universities.

Brazil Is a New Democracy

Brazil is a fairly new democracy. Therefore, it is still cutting its teeth and shaping its future as a democratic country. Barros states that "In Brazil, it is evident the perception that the democratic rule of law, far to achieve stability, suffers from certain legal and institutional anachronisms, notary and clientelistic practices, in addition to entrenched interference in their education process historical, social, political, cultural and economic."[154]

Brazilians are still discovering the value of making their voices heard and the richness of truly democratic institutions. From 2011 through 2016, thousands of people protested governmental acts in different parts of the country. Reasons for protesting ranged from complaints about raising the price of public transportation to corruption and discontentment with the government of the former president, Dilma Rousseff.[155] However, Brazil still needs to make its processes more transparent and encourage public participation on legislative proposals and administrative decisions through social media and other means. For example, in February 2016, CONAMA posted a draft resolution on its website to collect the public's comments. The draft resolution proposed changes in federal environmental permitting procedures and was available for consultation for ten days; however, the consultation period occurred during Brazilian Carnival, a long holiday period when most Brazilians are on vacation and traveling. Therefore, the *Ministério Público* requested cancellation of this public consultation[156] and instead proposed that the matter be discussed in public meetings. This example demonstrates a lack of interest from the government in consulting Brazilians on a very important issue: the environmental permitting procedure. The government made the draft resolution available for a consultation that would not reflect the population's input. However, it gave the "appearance" of a democratic process. Such attitude leads to the next section, where corruption, economic interests, and political will are powerful forces that often try to masquerade as public participation in EIAs.

Corruption, Economic Interests, and Political Will

The most problematic and difficult barriers to strong public participation are corruption, the power of economic interest, and the political forces behind projects' decision-making. Brazilian development model focuses on economic growth at almost any cost. Several examples demonstrate this fact, but a particularly useful illustration is the Brazilian National Development Bank (in Portuguese *Banco Nacional de Desenvolvimento Econômico e Social* [BNDES]), whose public domestic funds have financed hydroelectric dams in Peru, Ecuador, and Nicaragua. Little information is available on these investments, which BNDES claims are contracts under confidentiality clauses.[157] Additionally, in spite of people's rejection, legal actions, and research proving the social and environmental negative impacts of dams, there are twenty-six hydropower plants under construction and already implemented in the Amazon rainforest.[158]

Corruption plays an important role in this scenario of lack of interest in truly taking into consideration people's views and wishes. According to the 2016 Economic Forum Competitiveness rankings, in relation to public institutions, ethics and corruption, and public trust in politicians, Brazil occupies the 138th position among the 140 countries analyzed.[159] From 2015 to 2016, corruption scandals—such as the Belo Monte Dam questioned in court due to allegations of involvement of contractors in political corruption schemes[160]—dropped Brazil's position by eighteen places.[161]

Decision-making in EIA procedures is influenced by political interests. Ferraz points out that the number environmental permits issued to projects rises during election periods in the state of São Paulo.[162] For instance, in years in which there are mayoral elections, there is more authorization to initiate project implementation in municipalities where the mayor is from the same party as the governor.[163] Further, when there are gubernatorial elections, the number of environmental permits rises in the municipalities where the governor has statistically more chances to win elections.[164]

The decision to authorize a project should take into consideration the findings of the EIA; however, it is a discretionary decision.[165] In other words, the environmental agency may authorize a project that proves deleterious to the environment or to people's lives. In many instances, this is done under the excuse of public domain or public interest. However, practice has demonstrated that many projects claiming benefits for Brazilians do not actually develop the region where the project is installed. For instance, during a field visit to Altamira, in Pará State, Brazil—the city most affected by Belo Monte Dam implementation—it was clear that the city did not improve due to the project. Several interviews with the population confirmed this impression.

The best way to explain political forces and power of economic interests behind the approval of EIAs and implementation of development projects in Brazil is through examples. In these examples, the effects of minimal community engagement and appropriate public participation during the EIA procedure are evidenced. The implementation of the world's third largest hydroelectric dam, the Belo Monte Dam, was delayed by approximately 465 days. The cost of implementation increased from US$4.75 billion to over US$7.5 billion dollars (estimated as of 2013),[166] which the consortium blames on several work stoppages, strikes, and legal actions, including some requesting prior informed consent and the realization of public hearings.

The Volta Grande Project, a mining project owned by Belo Sun Mining, is another example of lengthy delays in the environmental permitting procedure associated with extra costs. This gold mining project is currently engaged in the Environmental Permitting Procedure. The company filed the request for an environmental permit in 2012. According to its own timeline, activities were to start in the beginning of 2015. Yet, many aspects are delaying the approval of this project including questions on access to information and public participation in its EIA procedure. As of 2019, the project has not started its operations. News of court suspension of the environmental permitting procedure seriously affected prices of Belo Sun's shares in the market.[167] These are practical examples where, in general, delays result from lack of political will to carry out truly public participation measures.

CONCLUSION

As the world was advancing in the environmental arena, Brazil—along with other developing countries—was resisting environmental regulations, alleging its need to grow economically and the unfairness of constraining its development, since developed countries had already played a significant role in "damaging" the environment, suggesting an intent to keep underdeveloped countries underdeveloped. Then, the EIA surged to manage the chaos of aggressive economic development and consequent high levels of pollution. The enactment of a regulation on EIA was a mix of external and internal pressure to adopt a mechanism able to orient decision-making of potentially harmful projects.

EIA first appeared in the legal sphere with limited scope under Law No. 6803/80, which established Basic Guidelines for Industrial Zoning in Critically Polluted Areas. Afterward, the National Environmental Policy recognized EIA as a tool of the environmental permitting procedure. In 1988, the EIA requirement prior to approval of projects was elevated to a constitutional mandate through the promulgation of the Federal Constitution.

CONAMA Resolution No. 001/86 is the main federal regulation on EIA. CONAMA Resolution No. 237/97, although considered unconstitutional by many scholars, was the only regulation on environmental competence to review EIAs in the country, and it raised many questions due to lack of clarity. Twenty years after the promulgation of the Federal Constitution, Congress enacted Law No. 140/11 to regulate environmental competences of the union, states, and municipalities. However, the law is still unclear. Moreover, it contains a provision that breaches the federalist system by allowing states' environmental councils to determine what should be considered local impacts in order to establish the competence of municipalities to conduct environmental permits. Such a decision does not consider the participation of municipalities.

In sum, the legal historical background demonstrates that EIA regulations present shortcomings, ranging from the definition of Environmental Impact Studies to more complex issues, such as the competence to review EIAs. Furthermore, the lack of participation from municipalities in decisions that affect their competence to conduct environmental permits clearly demonstrates an antidemocratic legal provision, which does not respect the country's constitutional political organization.

The implementation of Environmental Permitting Procedure has many shortcomings, such as poor quality of studies and lack of public participation. Criticisms of the fact that the procedure is long because of a great number of legal actions and excessive intervention of the *Ministério Público* does not stand because these are results of a defective procedure. Therefore, *Ministério Público* only intervenes when there are problems with the procedure and EISs presented. Therefore, it is necessary to correct these problems at their sources. Increased social control over EIAs, improved transparency, easy access to environmental assessments, and an increase in meaningful public participation are measures capable of fixing, in great part, some of the problems that result in legal actions filed by *Ministério Público*.

Public participation in Brazilian EIAs remains weak for several reasons. First, Brazilians are not incentivized to participate in administrative and legislative decisions and/or procedures leading to a lack of interest in participating. Second, as a recent democracy, Brazil is still shaping its institutions. Although these are relevant reasons, the most serious—and difficult to tackle—reasons are corruption and political will, which are driven by economic interests.

A more transparent EIA procedure that provides meaningful public participation, based on engagement of communities through education and construction of mutual trust, is necessary; eventually leading to consensus or quasi-consensus in the authorization of development projects. Furthermore, it would enhance social control over government's environmental and economic decisions.

Brazilian EIA legislation is reasonably well-written (and sometimes even more advanced than other nations' legislation). However, modernization and amendments are necessary to place public participation at the early stages of the EIA procedure as well as to provide measures capable of enhancing the quality of people's participation.

Understanding the background of Brazilian EIA regulations and its relationship with the environmental permitting procedure, the next chapter describes the case study—the Volta Grande Project—analyzing the main weaknesses of EIA procedures in light of the current Brazilian EIA norms and interrelated legislations.

NOTES

1. Paulo Cesar Gonçalves Egler, "Improving the Environmental Impact Assessment Process in Brazil" (PhD diss., School of Environmental Sciences University of East Anglia, 1998), 236, http://busca.ibict.br/SearchBDTD/search.do?command=search&q=+instituicao_defesa:%22University%20of%20East%20Anglia%22.

2. Carlos Alberto Giannasi, "A Doutrina de Segurança Nacional e o Milagre Econômico" (The National Security Doctrine and the Economic Miracle) (PhD diss., São Paulo University, 2011), 8, http://www.teses.usp.br/teses/disponiveis/8/8137/tde-30092011-170055/pt-br.php.

3. Egler, "Improving the Environmental Impact Assessment Process in Brazil," 236.

4. Aline Borges do Carmo and Alessandro Soares da Silva, "Licenciamento Ambiental Federal no Brasil: Perspectiva Histórica, Poder e Tomada de Decisão em um Campo em Tensão" (Federal Environmental Permitting in Brazil: Historical Perspective, Power and Decision-making in a Field Experiencing Tension), *Revista Franco-brasileira de Geografia* 19 (2013).

5. *Relatório da Delegação do Brasil a Conferência das Nações Unidas sobre o Meio Ambiente—Volume II* (Brazilian Delegation's Report to the United Nations Conference on the Human Environment—Volume II) (1972), https://cetesb.sp.gov.br/proclima/wp-content/uploads/sites/36/2013/12/estocolmo_72_Volume_II.pdf.

6. Interesting fact is that fire in the Amazon rainforest is a concern that balanced relations between Brazil, France, Germany, and Norway, many years later, in 2019. "Amazon Fires: Brazil Threatened over EU Trade Deal," *BBC News*, August 23, 2019, https://www.bbc.com/news/world-latin-america-49450495; "Norway Halts Amazon Fund Donation in Dispute with Brazil," *The Guardian*, August 16, 2019, https://www.theguardian.com/world/2019/aug/16/norway-halts-amazon-fund-donation-dispute-brazil-deforestation-jair-bolsonaro.

7. Egler, "Improving the Environmental Impact Assessment Process in Brazil," 236.

8. Ibid., 241, 243.

9. Bertha K. Becker and Claudio A. G. Egler, *Brazil: A New Regional Power in the World Economy* (Cambridge: University Press, 1992), 95–96.

10. Egler, "Improving the Environmental Impact Assessment Process in Brazil," 237.

11. Ibid., 247.

12. Ibid., 251.

13. Ibid., 254.

14. Luis Enrique Sanchez, *Avaliação de Impacto Ambiental: Conceitos e Métodos* (Environmental Impact Assessment Concepts and Methodologies), 2nd ed. (São Paulo: Oficina de Textos 2013), chap. 2, p. 10, E-BOOK, https://ebooks.ofitexto.com.br/product/avaliao-de-impacto-ambiental-2-ed.

15. B. Pagnoccheschi and M. Bernado, "Política Ambiental no Brasil" (Environmental Politics in Brazil), in *Territorio, Ambiente e Políticas Públicas Espaciais*, ed. Marília Steinberger (Brasília: LGE Publisher, 2006), 101–123; Borges do Carmo and Soares da Silva, "Licenciamento Ambiental Federal no Brasil," 9.

16. Borges do Carmo and Soares da Silva, " Licenciamento Ambiental Federal no Brasil," 9.

17. Antonio Herman de Vasconcellos Benjamin, "Os Principios do Estudo de Impacto Ambiental como Limites da Discricionariedade Administrativa" (The Principles of the Environmental Impact Assessment as Limits of the Administrative Discretion), *Review Forense* 317 (1992): 25.

18. Ibid.

19. See Clovis Cavalcanti, "Economic Growth and Environmental Protection in Brazil: An Unfavourable Trade-off," in *Environmental Governance and Decentralization*, eds. Albert Breton et al. (Cheltenham: Elgar Publishing, 2007), 56.

20. Decreto 99274, de 6 de Junho de 1990, art. 17, Diário Oficial da União (D.O.U.) de 07.06.1990 (Braz.). Article 7, II of Decree 99274/90 states CONAMA's responsibilities.

21. In some instances, EIA and EIS are used interchangeadbly in this book. As previously explained while EIA is the procedure EIS is the report that reflects the findings of the procedure. It is important to note that when this book refers to poor quality of EIAs/EISs it means, among other things, that the collection of data and methodoly used does not reflect best practice; the EIA/EIS is incomplete and/or does not follow the TOR previously agreed upon.

22. Édis Milaré, *Direito do Ambiente: A Gestão Ambiental em Foco* (Environmental Law: Environmental Management in Focus), 7th ed. (São Paulo: Editora Revista dos Tribunais, 2011), 467.

23. Antonio F. G. Beltrão, *Aspectos Jurídicos do Estudo de Impacto Ambiental (EIA)* (Legal Aspects of Environmental Impact Assessment [EIA]) (São Paulo: MP Editora, 2008), 34.

24. Milaré, *Direito do Ambiente*, 466.

25. Ibid.

26. Beltrão, *Aspectos Jurídicos*, 35.

27. Ibid., 34.

28. Ibid., 34; Milaré, *Direito do Ambiente*, 466.

29. Ibid., 467.

30. Ibid.

31. Resolução CONAMA No. 009, de 03 de Dezembro de 1987, art. 2, Diário Oficial da União (D.O.U.) de 05.07.1990 (Braz.).

32. Ibid., art. 2 para. 2.

33. Beltrão, *Aspectos Jurídicos*, 27.

34. Celso Antonio Pacheco Fiorillo et al., *Licenciamento Ambiental* (Environmental Licensing) (Brazil: Saraiva, 2011).

35. Ibid., 60, 61; Tiago do Amaral Rocha and Mariana Oliveira Barreiros de Queiroz, "O Meio Ambiente como Direito Fundamental da Pessoa Humana" (The Environment as a Fundamental Right), *Revista Âmbito Jurídico* 95 (2011), http://ambito-juridico.com.br/site/?artigo_id=10795&n_link=revista_artigos_leitura.

36. Consuelo Yatsuda Moromizato Yoshida, "A Experiencia do Licenciamento Ambiental do Rodoanel Mario Covas e os Avancos no Trato das Questoes Federativas, Institucionais e Ambientais" (The Experience of the Environmental Licensing of the Mario Covas Beltway and the Progress Regarding the Federative, Institutional and Environmental Matters), in 1 *Paisagem, Natureza e Direito* (Landscape, Nature and Law), ed. Antonio Herman Benjamin (2005), 231, 236.

37. Constituição Federal (C.F.) (Constitution) art. 225, §1, IV (Braz.).

38. Milaré, *Direito do Ambiente*, 470.

39. Resolução CONAMA No. 237, de 19 de Dezembro de 1997, art.3, Diário Oficial da União (D.O.U.) de 22.12.1997 (Braz.).

40. Milaré, *Direito do Ambiente*, 472.

41. José Fernando Vidal de Souza & Erika Von Zuben, "O Licenciamento Ambiental e a Lei Complementar n.140/2011" (The Environmental Licensing and the Federal Law No. 140/2011), *Caderno de Direito* 12 (2012), 11.

42. Sidney Guerra, "Competência Ambiental a Luz da Lei Complementar No. 140/2011" (Environmental Competence in Light of the Federal Law No. 140/2011), *Revista do Programa de Pós-Graduacao em Direito da UFC* 32 (2012), 125, http://www.periodicos.ufc.br/index.php/nomos/article/viewFile/377/359.

43. Ibid.

44. Do Valor, *Obras de Pólo Petroquimico do Rio São Interrompidas por Anulação de Licenças* (Licenses Authorizing Implementation of the Petrochemical of Rio are Canceled), *Folha de São Paulo*, May 15, 2013, 9:24 a.m., http://www1.folha.uol.com.br/mercado/2013/05/1278907-obras-do-comperj-da-petrobras-sao-interrompidas-por-anulacao-de-licencas.shtml.

45. Vidal de Souza and Von Zuben, "O Licenciamento Ambiental e a Lei Complementar," 11.

46. Guerra, "Competência Ambiental," 125.

47. Ibid.

48. Lei Complementar No. 140/11, de 8 de Dezembro de 2011, art. 9, XIV "a," Diário Oficial da União (D.O.U.) de 9.12.2011 retificado em 12.12.2011 (Braz.); Lucas Azevedo de Lima, "Uma Análise sobre a Responsabilidade do Conselho Estadual de Determinar a Tipologia para o Município Promover o Licenciamento das Atividades de Impacto Ambiental" (An Analysis of the Responsibility of the State Council to Determine the Competence of the Municipality to Conduct Environmental Licensing of Activities that Causes Environmental Impacts), *Interfaces Científicas* 1 (2013), 49.

49. Azevedo de Lima, "Uma Análise sobre a Responsabilidade do Conselho Estadual," 49.

50. Vidal de Souza and Von Zuben, "O Licenciamento Ambiental e a Lei Complementar," 11.

51. "Cidades, Maranhão, Belágua" (Cities, Maranhão, Belágua), Instituto Brasileiro de Geografia e Estatistica (IBGE), last visited May 25, 2015, http://cidades.ibge.gov.br/xtras/temas.php?lang=&codmun=210173&idtema=16&search=||s%EDntese-das-informa%E7%F5es.

52. Sidney Guerra, "Competência Ambiental a Luz da Lei Complementar No. 140/2011" (Environmental Competence in Light of the Federal Law No. 140/2011), *Revista do Programa de Pós-Graduacao em Direito da UFC* 32 (2012), 125, http://www.periodicos.ufc.br/index.php/nomos/article/viewFile/377/359.

53. John Glasson and Nemesio Neves B. Salvador, "EIA in Brazil: A Procedure-practice Gap. A Comparative Study with Reference to the European Union, and Especially the UK," *Environmental Impact Assessment Review* 20, no. 2 (April 2000), 191.

54. John Glasson et al., *Introduction to Environmental Impact Assessment*, 3rd ed. (London: Routledge, 2005), 36.

55. Ibid.

56. Ibid.

57. John W. Stampe, "Lessons Learned from Environmental Impact Assessments: A Look at Two Widely Different Approaches—The USA and Thailand," *Journal Transdisciplinary Environmental Studies* 8, no. 1 (2009).

58. Resolução CONAMA No. 001, de 23 de Janeiro de 1986, Diário Oficial da União (D.O.U.) de 17.02.1986 (Braz.).

59. Ibid., art. 10.

60. World Bank, *Environmental Licensing for Hydroelectric Projects in Brazil: A Contribution to the Debate* (March 2008), http://siteresources.worldbank.org/EXTWAT/Resources/4602122-1214578930250/Summary_Report.pdf; Fiorillo et al., *Licenciamento Ambiental*, 81.

61. Resolução CONAMA No. 237/97, art. 8(I).

62. Ibid., art. 8(II).

63. Ibid., art. 8(III).

64. See World Bank, *Environmental Licensing for Hydroelectric Projects in Brazil*.

65. Ibid.

66. Raul Silva Telles do Valle, "Sociedade Civil e Gestão Ambiental no Brasil: uma Análise da Implementação do Direito a Participação em nossa Legislação" (Civil Society and Enviromental Management in Brazil: An Analysis of the Implementation of the Right to Participate in our Legislation) (Master's thesis, Universidade de São Paulo, 2002) (on file with author).

67. Resolução CONAMA No. 237, de 19 de dezembro de 1997, Diário Oficial da União (D.O.U.), de 22.12.1997 (Braz.).

68. Antonio Inage de Assis Oliveira, *O Licenciamento Ambiental* (Environmental Licensing), ed. Julio Igliori (1998), 121.

69. Constituição Federal (C.F.) (Constitution) art. 225, § 1, IV (Braz.); Paulo Affonso Leme Machado, "Legislação de Avaliação de Impactos Ambientais no Brasil" (Legislation on Environmental Impact Assessment in Brazil), in *Manual de Avaliação de Impactos Ambientais* (Environmental Impact Assessment Guide), ed. Secretaria Especial do Meio Ambiente do Paraná (SEMA-PR) (Curitiba: Surehma, 1992), 640, as cited in Oliveira, *O Licenciamento Ambiental* (Environmental Licensing), 121.

70. Glasson et al., *Introduction to Environmental Impact Assessment*, 4.

71. Christopher Wood, *Environmental Impact Assessment: A Comparative Review*, 2nd ed. (London: Routledge, 2003).

72. Glasson et al., *Introduction to Environmental Impact Assessment*, 5.

73. Ibid., 4.

74. Ibid., 5.

75. Council on Environmental Quality (CEQ) Regulations for Implementing the Procedural Provisions of the National Environmental Policy Act, 40 C.F.R. pts. 1500–1508 (2015).

76. Neil Craik, *The International Law of Environmental Impact Assessment: Process, Substance and Integration* (Cambridge: Cambridge University Press, 2008), 42.

77. Resolução CONAMA No. 001/86, art. 11 para. 2.

78. *Ministério Público* is comprised of Brazilian public prosecutors. It is an independent and permanent institution, whose objective is to defend the juridical order, the democratic regime, and the social and indispensable rights. Article 127 of the Brazilian Federal Constitution determines the principles of the *Ministério Público*.

79. Resolução CONAMA No. 237/97, art. 10.

80. Ministério do Meio Ambiente et al., *Guia de Procedimentos do Licenciamento Ambiental Federal* (Federal Environmental Licensing Guide) (2002), http://www.mma.gov.br/estruturas/sqa_pnla/_arquivos/Procedimentos.pdf.

81. Ibid.

82. Resolução CONAMA No. 001/86.

83. Ministério do Meio Ambiente et al., *Guia de Procedimentos do Licenciamento Ambiental Federal.*

84. Gisela Foratini, Environmental Licensing Director, Brazilian Institute of Environment and Renewable Natural Resources, Brazil, *Licenciamento Ambiental Federal. Avanços e Desafios* (Environmental Permitting Procedure. Progress and Challenges) (Apr. 30, 2013), British Chamber of Commerce in Brazil, https://sitefiespstorage.blob.core.windows.net/uploads/2013/05/8.30-GISELA-Forattini.pptx.

85. Impact Assessment Act S.C. 2019, c. 28, s. 1, 28 (2), https://laws.justice.gc.ca/PDF/I-2.75.pdf.

86. Ministério Público de Minas Gerais, *Guia Técnico para Atuação do Ministério Público no licenciamento Ambiental de Atividade de Mineração* (Technical Guide for the Role of Public Prosecutors at the Environmental Permitting of Mining Activities), (Minas Gerais: Rev. do Ministério Público do Estado de Minas Gerais, 2012), www.mp.mg.gov.br/mpmgjuridico.

87. Ibid.

88. Ibid.

89. Ibid.

90. Ministério do Meio Ambiente, *Manual de Normas e Procedimentos para Licenciamento Ambiental no Setor de Extração Mineral* (Handbook of Environmental Licensing Rules and Procedures of the Mineral Extractive Sector) (2001), http://www.mma.gov.br/estruturas/sqa_pnla/_arquivos/MANUAL_mineracao.pdf.

91. Ibid.

92. Glasson et al., *Introduction to Environmental Impact Assessment*, 15; A. John Sinclair and Patricia Fitzpatrick, "Provisions for More Meaningful Public Participation Still Elusive in Proposed Canadian EIA Bill," *Impact Assessment & Project Appraisal* 20, no. 3 (2002): 161, 163; Nichola Hartley and Christopher Wood, "Public Participation in Environmental Impact Assessment—Implementing the Aarhus Convention," *Environmental Impact Assessment Review* 25, no. 4 (May 2005): 319.

93. Philippe Sands et al., *Principles of International Environmental Law*, 3rd ed. (Cambridge: Cambridge University Press, 2012).

94. Paulo Affonso Leme Machado, *Direito Ambiental Brasileiro* (Brazilian Environmental Law), 21st ed. (São Paulo: Malheiroso Editores, 2013).

95. Rachel Davis and Daniel M. Franks, *Costs of Company-Community Conflict in the Extractive Sector* (Cambridge, MA: Harvard Kennedy School, 2014), https://www.csrm.uq.edu.au/media/docs/603/Costs_of_Conflict_Davis-Franks.pdf.

96. Flavia Milano and Andrea Sanhueza, *Public Consultations with Civil Society: Guidelines for Public and Private Executing Agencies*, March 2016, https://publications.iadb.org/handle/11319/7499.

97. Brian Ganson and Achim Wennmann, *Business and Conflict in Fragile States: The Case for Pragmatic Solutions* (London: Routledge, 2016), 22.

98. Emily Greenspan, Free, *Prior and Informed Consent in Africa* (Oxfam America, 2014), www.oxfamamerica.org/publications/fpic-in-africa.

99. Witold J. Henisz et al., "Spinning Gold: The Financial Returns to Stakeholder Engagement," *Strategic Management Journal* 35, no. 12 (2014): 1727.

100. International Finance Corporation, *Policy on Social and Environmental Sustainability*, January 11, 2012, 6, https://www.ifc.org/wps/wcm/connect/7141585d-c6fa-490b-a812-2ba87245115b/SP_English_2012.pdf?MOD=AJPERES&CVID=kiIrw0g.

101. Rodrigo Baptista, "Avança Projeto que Acelera Liberação de Licenças Ambientais para 'Obras Estratégicas'" (Proposal of Granting Faster Environmental Licenses Authorizing Implementation of Governmental Strategic Projects Moves Forward), *Agencia Senado*, November 26, 2015 10:26 a.m., http://www12.senado.leg.br/noticias/materias/2015/11/25/avanca-projeto-que-acelera-liberacao-de-licencas-ambientais-para-obras-estrategicas.

102. Ibid.

103. Fiorillo et al., *Licenciamento Ambiental*, 183.

104. Ibid.

105. The timeline is understood here as the moment that the project proponent requests authorization to the permitting agency and receives approval. World Bank, *Environmental Licensing for Hydroelectric Projects in Brazil.*

106. Gary Middle and Isaac Middle, "The Inefficiency of Environmental Impact Assessment: Reality or Myth?," *Impact Assessment & Project Appraisal* 28, no. 2 (2012): 159–61.

107. Atif Ansar et al., "Should We Build More Large Dams? The Actual Costs of Hydropower Megaproject Development," *Energy Policy* 69 (June 2014): 43, 45, 48.

108. This includes the moment of filing the request for a preliminary license up until operation of the project.

109. World Bank, *Environmental Licensing for Hydroelectric Projects in Brazil.*

110. Ibid.

111. Klemens Laschefski, "Comunidades Tradicionais e Licenciamento Ambiental 'Entraves' para o Desenvolvimento" (Traditional Communities and Environmental Licensing: Are they Barriers to Development?), (PowerPoint presentation, Environmental Licensing: Reality and Perspectives, Brasília, Nov. 5, 2015), http://www.mpf.mp.br/atuacao-tematica/ccr4/dados-da-atuacao/eventos/Seminarios/KlemensLaschefski.pdf.

112. Ibid.

113. Tribunal de Contas da União, *Acórdão No. 3413 Plenário, Licenciamento das UHEs de Jirau e Santo Antônio* (Environmental Permitting of the Hydroelectric Dams Jirau and Santo Antônio) (December 5, 2012).

114. Tribunal de Contas da União, *Acórdão No. 2.212 Plenário, Relatório de Levantamento—Avaliação do IBAMA* (Report of Analysis of the Federal Environmental Agency—IBAMA) (September 23, 2009).

115. Rose Mirian Hofmann, *Gargalos do Licenciamento Ambiental Federal no Brasil* (Shortcomings of the Federal Environmental Permitting in Brazil) (2015), http://bd.camara.gov.br/bd/handle/bdcamara/24039.

116. World Bank, *Environmental Licensing for Hydroelectric Projects in Brazil.* Although this report was written prior to the law that establishes environmental competence (Law No. 140/2011), competence of environmental agencies (federal or state) is still very debatable.

117. McAllister states that

> The Brazilian *Ministério Público* consists of a state *Ministério Público* in each of Brazil's 26 states as well as a Federal *Ministério Público* with representatives in each state. The term *Ministério Público* has sometimes been translated into English as the "public prosecution." The most analogous institution in the United States is the Office of the Attorney General.

However, there are some differences between the United States Attorney General and the *Ministério Público. See* Lesley K. McAllister, "Environmental Advocacy Litigation in Brazil and the United States," *Journal of Comparative Law* 6, no. 2 (2011): 203.

118. Constituição Federal (C.F.) (Constitution) art. 127 (Braz.).

119. McAllister, "Environmental Advocacy Litigation," 203.

120. The following actors have legitimacy to file a public civil action: the *Ministério Público*, public defenders, union, states, Federal District and municipalities, local authorities, public enterprises, foundation or joint capital company, and associations which have been operating for at least one year and have as mandate environmental and social causes. The public civil action is the proper action to argue environmental issues. Lei No. 7.347, de 24 de Julho de 1985, Diário Oficial da União (D.O.U.) de 25.07.1985 (Braz.).

121. McAllister, "Environmental Advocacy Litigation," 203.

122. World Bank, *Environmental Licensing for Hydroelectric Projects in Brazil.*

123. Hofmann, *Gargalos do Licenciamento Ambiental Federal no Brasil.*

124. Ibid.

125. World Bank, *Environmental Licensing for Hydroelectric Projects in Brazil.*

126. El-Sayed A. Badr, "Evaluation of the Environmental Impact Assessment System in Egypt," *Impact Assessment & Project Appraisal* 27, no. 3 (2012): 193.

127. Fiorillo points out that public participation in the United States is well-oriented. See Fiorillo et al., *Licenciamento Ambiental*, 173.

128. In *Robertson v. Methow Valley Citizens Council*, the EIA estimated that the construction of a sky resort in Okanogan National Forest, in Okanogan County, Washington, would impact 30,000 deer in the region. The Supreme Court stated that NEPA § 101, which sets the EIA procedure, is a set of "action forcing" procedures that require that agencies take a hard look at environmental consequences. Therefore, NEPA merely prohibits uninformed rather than unwise agency actions. *See Robertson v. Methow Valley Citizens Council*, 490 U.S. 332 (1989).

129. Ley No. 25.675, Nov. 27, 2002 (Arg.).

130. Iara Verocai, "O Licenciamento Ambiental em Outros Paises" (Environmental Licensing in Other Countries) (Notes from Conference on Permitting Procedure,

Environmental Protection and Development, July 29, 2004), http://www.mma.gov.br/estruturas/DAI/_arquivos/iaraverocai2.pdf.

131. Ibid.

132. Milano and Sanhueza, *Public Consultations with Civil Society*.

133. Servicio de Evaluación Ambiental, Qué es la Participación Ciudadana en el SEIA? (What is Public Participation in the Environmental Impact Assessment Service?) (last accessed December 15, 2019), https://sea.gob.cl/participacion-ciudadana-y-consulta-indigena/que-es-la-participacion-ciudadana-en-el-seia. Decreto No. 40, Octubre 30, 2012, Diario Oficial (D.O.) (Chile).

134. Milano and Sanhueza, *Public Consultations with Civil Society*.

135. Ibid.

136. Milano and Sanhueza, *Public Consultations with Civil Society*.

137. Ibid.

138. Servicio de Evaluación Ambiental, Qué es la Participación Ciudadana en el SEIA? (What is Public Participation in the Environmental Impact Assessment Service?) (last accessed December 15, 2019), https://sea.gob.cl/participacion-ciudadana-y-consulta-indigena/que-es-la-participacion-ciudadana-en-el-seia.

139. Ednaldo Cândido Rocha et al., "Avaliação de Impactos Ambientais nos Países do Mercosul" (Environmental Impact Assessment in the Mercosur Countries), *Ambiente & Sociedade* 8, no. 2 (2005): 147.

140. The study was completed between May and September 2014. "Brazil," Environmental Democracy Index, last visited September 15, 2019, http://environmentaldemocracyindex.org/country/bra.

141. "Egypt," Netherlands Commission for Environmental Assessment, last updated August 28, 2019, http://www.eia.nl/en/countries/af/egypt/eia.

142. Lucivaldo Barros Vasconcelos, "O Estado (In) Transparente: Limites do Direito a Informação Socioambiental no Brasil," (The Lack of Transparency of the State: Limits of Social and Environmental Information in Brazil) (PhD dissertation, Universidade de Brasília, 2008), 309, 336, http://repositorio.unb.br/handle/10482/4167?mode=full. Barros explains that lack of information is a major problem and it is a barrier to participation.

143. "Intelligence Unit: Democracy Index 2018: Me too? Political Participation, Protest and Democracy 20," The Economist, last accessed September 15, 2019, http://www.eiu.com/Handlers/WhitepaperHandler.ashx?fi=Democracy_Index_2018.pdf&mode=wp&campaignid=Democracy2018.

144. Ibid.

145. "Brazil," OECD Better Life Index, last accessed September 15, 2019, http://www.oecdbetterlifeindex.org/countries/brazil/.

146. "Referendo" (Referendum), Tribunal Superior Eleitoral (Superior Electoral Tribunal), last accessed September 15, 2019, http://www.tse.jus.br/eleitor/glossario/termos/referendo.

147. "UK Referendum," BBC News, last accessed September 15, 2019, https://www.bbc.com/news/politics/eu_referendum/results.

148. Ariane dos Santos Hoppe et al., "Participação Popular no Sistema Único de Saúde: Olhar de Usuários de Serviços de Saúde" (Public Participation in the

Unified National Health Care System: Health Care Users' View), *Cinergis* 18, no. 1 (December 2017): 4.

149. Diana Meirelles da Motta and Bolívar Pêgo, eds., *Licenciamento Ambiental para o Desenvolvimento Urbano: Avaliação de Instrumentos e Procedimentos* (Environmental Permitting Procedure for the Urban Development. Evaluation of Tools and Precedures) (Rio de Janeiro: IPEA, 2013), 188; Fernando Marcelo, "Macaé: Realizada Audiência Pública das Termelétricas Jaci e Tupã," *Pre-Sal Info*, July 17, 2019, https://www.pre-sal.info/2019/07/17/macae-realizada-audiencia-publica-das-termeletricas-jaci-e-tupa/.

150. Da Motta and Pêgo, *Licenciamento Ambiental*, 188.

151. Ibid.

152. This section expresses the author's view of reasons why public participation is weak in Brazilian EIAs. It does not intend, by any means, to carry out a deep analysis of weak public participation in Brazil, from the sociological point of view.

153. *See generally* Barros, "O Estado (In) Transparente."

154. Ibid.

155. "Todos os Estados e o DF tem Protestos Contra o Governo Dilma" (Protests are Happening in All States and the Capital Against Dilma), G1, August 17, 2016, 6:47 a.m., http://g1.globo.com/politica/noticia/2015/08/manifestantes-protestam-contra-o-governo-dilma-em-cidades-do-brasil.html.

156. "MPF Recomenda ao Conselho Nacional do Meio Ambiente Consulta à População sobre Licenciamento Ambiental" (Federal *Ministério Público* Recommends to the National Environmental Council Holding Public Consultation Regarding Environmental Licensing Proposal), Ministério Público Federal, March 7, 2016, 12:51 p.m., http://www.mpf.mp.br/pgr/noticias-pgr/mpf-recomenda-ao-conselho-nacional-do-meio-ambiente-consulta-a-populacao-sobre-licenciamento-ambiental.

157. Celio Bermann, "Governance of Infrastructure Investment in Brazil: Implications of Energy Related Mega Projects for Society, the Environment and Democracy," in *Infrastructure: For People or For Profit?* ed. Nancy Alexander (Heinrich Boell Foundation, 2014).

158. Ibid.

159. World Economic Forum, *The Global Competitiveness Report 2015–2016: Brazil* (2016), http://reports.weforum.org/global-competitiveness-report-2015-2016/economies/#economy=BRA.

160. *See* Miriam Leitao, "Delcidio Esclarece" [Delcidio, former Brazilian Senate, Clarifies], *O Globo* (blog), March 3, 2016, 8:30 a.m., http://blogs.oglobo.globo.com/miriam-leitao/post/delcidio-esclarece.html.

161. Ibid.

162. Claudio Ferraz, *Electoral Politics and Bureaucratic Discretion: Evidence from Environmental Licensing and Local Elections in Brazil* (unpublished working paper, January 2007), https://pdfs.semanticscholar.org/4c65/b4c514acbe53aa7bd24250d2d5c8e6cab4d2.pdf.

163. Ibid.

164. Ibid.

165. Fiorillo et al., *Licenciamento Ambiental*.

166. Renee Pereira, "Orçado em R$ 16 Bilhões, Custo da Usina de Belo Monte já Supera os R$ 30 Bilhões" (Belo Monte Dam Initial Budget of 16 Billion has already Surpassed 30 Billion Reais), *Estadão*, May 11, 2013 11:00 p.m., http://economia.estadao.com.br/noticias/geral,orcado-em-r-16-bilhoes-custo-da-usina-de-belo-monte-ja-supera-os-r-30-bilhoes,153398e.

167. TRF-1, AI No. 0011749-24.2015.4.01.0000/Pa Pará, Relator: Des. Jirair Aram Meguerian, 12.03.2015 (Braz.).

Part II

CASE STUDY IN THE BRAZILIAN AMAZON RAINFOREST

Chapter 3

Volta Grande Project

Belo Sun Mining Corporation, Ltd., is a Canada-based mineral exploration company.[1] Its Brazilian subsidiary, Belo Sun Mineração Ltda. (Belo Sun), is engaged in an environmental permitting procedure to install a gold mining project in Brazil: the Volta Grande Mining Project (Volta Grande Project or Project). The Volta Grande project is slated to be a Brazilian Amazon open-pit mining operation, located in the northern region of Brazil—the Senador José Porfírio municipality in the state of Pará. The project covers a total area of 2,356.41 hectares or 23.5641 square kilometers.[2]

LOCALIZATION OF THE VOLTA GRANDE PROJECT

Senador José Porfírio (SJP) is a 14,419.916 square kilometer municipality.[3] It has an estimated population of 11,658 people (2019 data) and the GDP per capita is R$10,272.06 (US$3,160), which is among the lowest in the state of Pará.[4] The level of education is low in SJP, where over 30 percent of its inhabitants are illiterate and labor is unqualified.[5] As of 2017, only 659 people or 5.8 percent of the population had jobs in the entire region.[6]

An important peculiarity about SJP is that its territory is discontinuous.[7] In other words, its territory is interrupted by another municipality (Vitoria do Xingu). This important geographic characteristic has consequences on participation in the EIA procedure for the Volta Grande Project.

SJP is 50 kilometers southeast of Altamira, which is the main municipality of an otherwise predominantly rural region.[8] Access to SJP is by way of small boats, known locally as *voadeiras*, or through a partially paved road.[9] Notably, the economies of Altamira and SJP are interrelated due to their proximities. For instance, selling agricultural goods, boat transportations, and gold negotiations from artisanal mining in SJP have impacted the market in both cities.[10] Coelho states that "artisanal mining is informal, does not generate direct taxes to the country; however, the money obtained [with artisanal mining] circulates in the city, moving local and regional economies."[11] In this region, artisanal mining has been happening for decades. Although sometimes illegal, the state has been absent in controlling this activity.

The project is planned for downstream of the Xingu River (on the right bank), namely Volta Grande do Xingu or Big Bend (Volta Grande).[12] The Xingu River is the second largest clearwater river system in South America. The Volta Grande is a 100-kilometer curve of the Xingu River, formed by bedrock and multiple flow path channels. Shawakushi et al. state that "[t]he tremendous size and morphological complexity of the rapids [flow] combined with a high variation of water level between the dry and wet seasons, make the Volta Grande a unique environment for Amazonian biodiversity."[13] The difficulty navigating the Volta Grande waters served as barriers to establish population on its margins during colonial times.[14] The Volta Grande is home to several ornamental fishes and fishing provides the prime source of protein for the riverine communities. The Volta Grande serves the basic needs as well as recreational and spiritual purposes for the people who live in its surroundings.[15]

The Brazilian Environmental Ministry points to the ecological relevance of the Volta Grande as a region of extremely high importance to the conservation of the biodiversity and priority areas for sustainable use[16] (figure 3.1).

The Volta Grande is occupied by traditional populations, which include indigenous and nonindigenous groups, such as riverine and artisanal miner communities. The EIS of Volta Grande asserts that "Indigenous Populations have been inhabiting the area for centuries."[17] Due to indigenous presence there are several archaeological sites in the Volta Grande. The EIS identified nine of them; however, it points out that this number can be higher. This means that further studies would be necessary to identify these areas.[18] The EIS states that currently the archaeological sites have been impacted by agriculture, cattle ranching, roads, and residential houses. Furthermore, current installations of the Volta Grande Project are located in one archaeological remain. Therefore, the EIS recognizes that actions to protect these sites are very important.[19]

Former artisanal gold mining in the area of the Volta Grande Project include the Ressaca, Ouro Verde, and Galo villages. These are federal public lands, which have been inhabited for decades by miners, agriculturists, indigenous peoples, fishermen, and riverbank communities. There are around 977 people living in the mentioned areas, which is home to five different communities (the biggest one, Ressaca village, has 452 residents). Moreover, the project is planned to be located close to four indigenous communities: the

Figure 3.1 The Volta Grande Area. *Source*: Photo taken by the author.

Paquiçamba, which is 12 kilometers away; Arara da Volta Grande, which is 16 kilometers away; Trincheira Bacaja, which is 39 kilometers away; as well as isolated indigenous communities living in the Ituna/Itata lands, which are 29 kilometers away.[20] The map below demonstrates the location of the project in relation to Paquiçamba and Arara da Volta Grande Indigenous Communities (figure 3.2).

The Volta Grande Project is planned for 10 kilometers away from the massive Belo Monte Dam.

THE BELO MONTE DAM

Belo Monte is a hydraulic power plant installed on the Xingu River. It is the world's third largest dam in installed generating capacity.[21] Belo Monte is actually composed of two reservoirs linked by an artificial canal that has 20 kilometers of extension. The main reservoir, in the Xingu River, is 359 square kilometers long. The second reservoir is 119 square kilometers long and is formed by 28 dikes and several canals.[22] When working at its total capacity, it is expected to generate 11,233 megawatts of electricity. However, the dam only works at its full capacity four months of the year (from February to May), when the Xingu River has its maximum flow. During the rest of the year, the turbines will be progressively

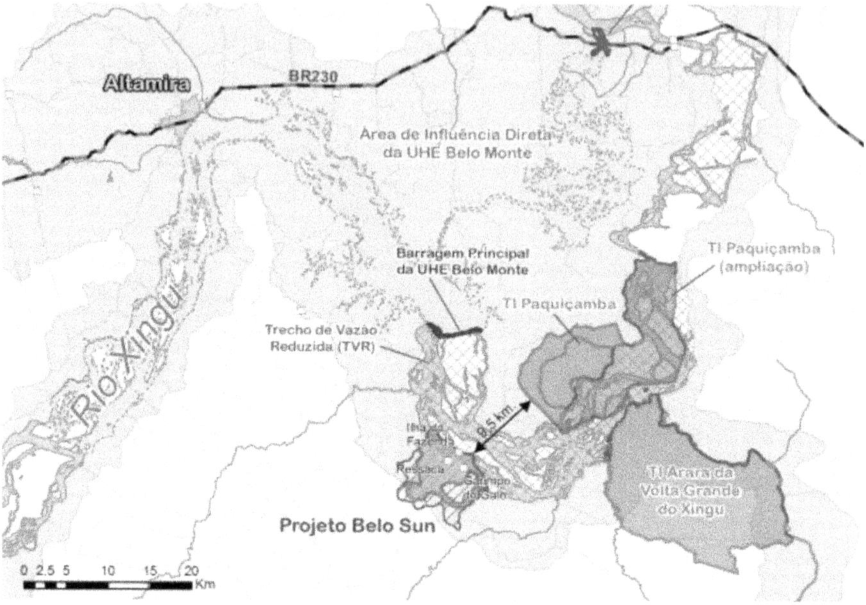

Figure 3.2 Map of the Volta Grande Project. The distance of 9.5 km from the Volta Grande Project to the Paquiçamba indigenous community was calculated by Instituto Socioambiental, considering an expansion of the Paquiçamba's land. Source: "Justiça Manda Suspender o Licenciamento da Belo Sun Mineração" [Court Suspends the Environmental Permitting Procedure of Belo Sun Mining], Instituto Socioambiental, November 21, 2013, https://www.socioambiental.org/pt-br/noticias-socioambientais/justica-manda-suspender-o-licenciamento-da-belo-sun-mineracao.

turned off and Belo Monte will generate only 4,571 megawatt. It is estimated that the dam works on just 40 percent of its installed capacity.[23]

Belo Monte is among the projects supported by the government through the Growth Acceleration Program (in Portuguese, *Programa de Aceleração do Crescimento*, or PAC).[24] It is being executed by Norte Energia, which is a public and private consortium. The construction of Belo Monte was very controversial. On April 1, 2011, the Inter-American Commission on Human Rights (IACHR) of the Organization of American States (OAS) advocated the suspension of Belo Monte environmental permitting procedure due to the harmful effects it would cause to indigenous communities living in the Xingu hydrographic basin, including the Arara da Volta Grande do Xingu, Juruna de Paquiçamba, isolated communities, among others. The IACHR required the realization of free, prior, and informed consent of indigenous communities as well as the adoption of measures capable of protecting the integrity of isolated indigenous communities and preventing dissemination of diseases among them in the Xingu hydrographic basin.[25] Former president Dilma Rousseff criticized the decision and reacted strongly. She suspended Brazilian contribution to the OAS (US$800,000) and withdrew

Brazil's ambassador to the OAS. Brazilian government representatives stated that the IACHR decision "threatened Brazilian Sovereignty."[26]

On July 29, 2011, the IACHR modified the provisional measures objective and determined that Brazil must adopt measures to guarantee the health and integrity of indigenous communities in the Xingu Basin. The commission stated the issue on indigenous peoples should be resolved under the merits of the case rather than discussed under provisional measures.[27]

The dam impacts the Volta Grande by diverting the natural flow of the river to artificial ditches and channels. Therefore the water flow that naturally occurs in this portion of the river is diverted to the main plant power station to generate energy.[28] As a condition of its installation permit Norte Energia must monitor the quality of water of the Volta Grande for six years, which might result in additional conditions on the Dam Operation permit.[29] According to IBAMA, the water quality in this specific area of the river during the installation and operation of Belo Monte is unpredictable. Therefore, it is necessary to monitor its quality and change dam operation by controlling the water flow, if necessary, to guarantee a healthy environment.[30]

The Environmental Impact Statement of the Belo Monte states that

> FUNAI [the Indigenous Affairs Agency] pointed to the necessity of more detailed studies in relation to Paquiçamba and Arara da Volta Grande do Xingu Indigenous Lands. . . . This is because [these lands] are in the Area of Direct Influence [of the Belo Monte Dam], they will be affected by the reduction of the Xingu river [in the Volta Grande area].[31]

Norte Energia also has an extensive set of obligations to improve the lives of the communities (indigenous and nonindigenous) living in the Volta Grande area in order to mitigate the impacts of its activities. Norte Energia has already started implementing some of the conditionalities of its permit, such as water and sewage installations.[32] In relation to Belo Monte's fulfillment of conditionalities prior to the issuance of the installation permit, Bratman states that

> Aside from such legal objections, however, the material power of the state is evident in the logistics of the project's continuation. The 2010 construction license for the dam was granted along with two long lists of preconditions, which purportedly needed to be met before the government would grant installation licenses. There were forty points on the list of environmental conditions, as well as thirty-eight indigenous conditions that needed to be met. . . . Construction continues despite very little progress in meeting the environmental or the indigenous conditions in the construction licenses, resulting in renewed court requests to halt the project.[33]

Importantly, in November 2015, Belo Monte was granted the operation permit despite the fact it did not comply with all conditions stated in the installation permit. However, in August 2016, the judiciary upheld Belo Monte's operation permit because the consortium did not meet a specific conditionality—implementation of sewage in Altamira[34]—which was ultimately postponed for years. Norte Energia (the consortium) alleged that its obligation was solely to build the sewage installation while the municipality had the obligation to operate the installation. Yet, the municipality pushed the obligation to Norte Energia. While there was no agreement of who would be responsible for providing sewage to the population—an obligation to be met before the subsequent permit would be granted—the permit was granted anyway.[35] On September 16, 2016, Belo Monte received authorization to operate the dam.[36]

In relation to mining activities in the Volta Grande, Belo Monte's EIS states that "Another impact analyzed was the possibility that, with the reduction of the flow during dry periods, it will increase artisanal mining in the Portion of Reduced flow [of the Xingu River—the Volta Grande]. This can increase conflicts with indigenous peoples as well as put pressure on their lands."[37] Furthermore, Norte Energia is responsible for monitoring mining activities in the area of influence of the dam. This measure is necessary to guarantee that other activities will not affect the structure of the dam.[38]

In sum, the area where Volta Grande Project is planned to be located has already been impacted by a massive dam. In particular, the Volta Grande will have its water flow greatly reduced, which opens the area to mining activities, as it was predicted by Belo Monte's EIS. As a consequence, Belo Monte has obligations in this area, such as monitoring mining activities and the quality of water. Thus, both projects are intrinsically related, since both will impact the Volta Grande area. Table 3.1 shows parallel actions of both projects: Belo Monte Dam and the Volta Grande.

THE GOLD DEVELOPMENT PROCESS AND INFRASTRUCTURE OF VOLTA GRANDE PROJECT

To extract ore, Belo Sun will perforate rocks utilizing tractors and explosives. Trucks and excavators will transport the material to the treatment plant where the gold will be separated. There will be two waste rock piles from the extraction of ore, which will be placed close to the pit.[39] Drainage systems will be installed in order to remove water in the area of the project. Additionally, cyanide will be used to treat the ore and a waste dam will contain the waste derived from the processing of gold.[40]

One of project's alternatives for waste dam was an area of 161.2 hectares with a reservoir of 25.5 million cubic meters of capacity, containing wastes—including cyanide.[41] Volta Grande infrastructure will consist of a

Table 3.1 Timeline of Belo Monte Dam and the Volta Grande Projects (2002–2018)

Year	*Belo Monte Dam*	*Volta Grande Mining*
2002	Project reinvigorated from the 1970s proposal; new viability studies conducted by Norte Energia.	
2007	Belo Monte Hydroelectric Project included in the Plan for Accelerated Growth (PAC I).	
2009	First environmental impact statement (EIS) is filed at IBAMA.	
2010	Preconstruction permit issued by the Federal Environmental Agency—IBAMA. The agency's head, Roberto Messias Franco, resigns over related differences of opinion.	Verena Mineração filed a draft TOR for the EIA of the Volta Grande mining project with the State Environmental Agency (SEMAS).
2011	Inter-American Human Rights Commission finds Brazil guilty of violations of FPIC. IBAMA's new head, Abelardo Bayma Azevedo, also resigns. Construction begins.	Verena Mineração was bought by Belo Sun and changed its corporate name to Belo Sun Mineração Ltda.
2012	Protests against dam continue, especially in conjunction with the international Rio+20 meeting. Legal requests filed in 2008–2010 remain largely unadjudicated.	Belo Sun files its EIS and its nontechnical summary.
2013	In Altamira, rates of murders double, traffic accidents increase by 144%, and sexual violence rates reach new highs. The population in the city doubles between 2010 and 2014. Twenty-five thousand urban residents are displaced from their homes and relocated.	The State Environmental Council voted in favor of granting the preconstruction permit. A court injunction suspended the environmental permit.
2014	Intermittent work stoppages due to worker conflicts and court injunctions which are later overruled.	Preconstruction permit was issued by SEMAS. A court injunction suspended the preconstruction permit.
2015	Operation permit issued.	
2016	Dam is inaugurated by President Dilma Rousseff; Corruption scandal testimonies indicate that between 1% and 2% of Belo Monte construction monies went into campaign kickbacks.	Brazilian National Council of Human Rights visited the area of the Volta Grande Project and heard local affected communities.
2017	Some of the initial preconditions from the permit remain unsatisfied, most notably the connection of homes in Altamira to the sewage system and the new hospital.	SEMAS and Belo Sun carried out workshops with local communities. SEMAS issued the installation permit on February 2.
2018		The Federal Court suspended the environmental permit alleging that the federal agency IBAMA would be responsible for conducting the permitting procedure of Belo Sun.

Source: Adapted from E. Bratman and C. Bena Dias, "Development Blind Spots and Environmental Impact Assessment: Tensions Between Policy, Law and Practice in Brazil's Xingu River Basin," *Environmental Impact Assessment Review* 70 (2018): 1–10.

laboratory, offices, warehouses, cafeteria, restrooms and changing rooms, explosive warehouses, medical station, a repair shop, a gas station, reception areas, parking places, workers' accommodation, and a transmission line of approximately 70 kilometers, which will bring 138 kilovolts of energy from Altamira City to the project. Sewage and water treatment stations will also be part of the project's new construction.[42]

The project will operate for twelve years. During its installation 2,100 people will work in the civil construction and will also carry out administrative and technical duties.[43] Belo Sun's EIAs estimates that the project will improve the quality of life for people in the region by creating employment, improving access roads, and paying taxes to the SJP municipality, the state of Pará, and the Federal Government, as well as recuperating the area, which was degraded by artisanal mining activities carried out in the region for decades.[44]

According to Brazilian legislation, gold companies shall pay 1 percent of net profit to the government in the form of royalties: 65 percent of royalties go to the municipality, 23 percent go to the state, and 12 percent is directed to the union.[45] Therefore, the municipality benefits directly with the payment of royalties.[46] If the project is approved, SJP will receive R$95 million (US$29.8 million) plus R$38 million of service tax (US$11.9 million); Pará State will receive R$17 million and the union R$8 million (US$5.3 million and 2.5 million respectively). According to Belo Sun, in its twelve years of operation, the company will pay R$158 million or US$49.5 million in taxes.[47]

The project will also have socio-environmental impacts. For instance, it will dramatically change the landscape of the Volta Grande through deforestation, movement of trucks, and dislodged rocks.[48] Furthermore, disturbances derived from explosions, and an increase of people in the area who will utilize the infrastructure of Altamira and SJP, will also generate negative impacts. According to the Volta Grande EIS Nontechnical Summary Report:

> In addition to the disruptions of affective, social, family and cultural ties that revolve around the mining, displacement process will cause interference, mainly in mining activity, affecting prices in the region and the jobs open for exploration of gold. In this situation, being resident or not the communities concerned, the shift will cause a very significant impact, since the inclusion of the project in that area, on the one hand, decreases the product offering in the regional market and, on the other hand, extinguish, for a moment, the main economic activity of most families of Ressaca Village, Garimpo do Galo and Ouro Verde.[49]

Although the company has plans to mitigate such impacts through environmental management, social communication, environmental education, health and safety programs, and communities' relocation, among others,[50] some impacts are irreparable, such as the loss of social and affective ties for people who live in the area, as mentioned in the EIS Nontechnical Summary report.

The Socioambiental Institute, an NGO closely monitoring the implementation of the Belo Monte Dam, points out the impacts of the Volta Grande Project on the indigenous communities. With the implementation of the project, these communities will suffer a significant increase of population in the area. During construction, more than 2,000 people will work in the Volta Grande, which will increase marine and road transit of nonindigenous individuals.[51] Furthermore, after construction is finished, employees might look for other forms of sustaining themselves and their families.

In the context of deforestation and illegal logging, which is already in place in the region, this project represents additional risks to indigenous people, who depend culturally and physically on the land and the Xingu River. Additionally, their relationship with the Xingu River differs from nonindigenous persons; for instance, the Jurunas, who live in the Paquiçamba indigenous lands, see themselves as "Lords of Rivers" as they possess great fishing and navigability skills. Any contamination of the waters of the river will affect their emotional and physical health as well as their food source. In addition, risks of accidents, such as the bursting of a waste dam, would be devastating for indigenous and nonindigenous communities who live in the Volta Grande.[52] Ultimately, these indigenous communities are within the directly affected area of the Belo Monte Dam; therefore, both projects will influence their way of living.

As both projects are located in close proximity (10 km apart from each other) and both will impact the Volta Grande area, a comprehensive study of cumulative and synergic impacts should have been carried out in order to assess the viability of both projects. However, Belo Sun states that the Volta Grande Project will not have great influence on the water flow of the Volta Grande area and therefore the Volta Grande Project will have no influence on the Belo Monte Dam.[53] Nevertheless, as of this writing, Belo Sun has not presented robust studies of cumulative impacts of installation and operation of both projects.

THE VOLTA GRANDE PROJECT ENVIRONMENTAL PERMITTING PROCEDURE

On October 10, 2010, Verena Mineração Ltda. (which changed its name to Belo Sun Mineração Ltda. in November 2011) filed a consultation letter and a draft TOR for the development of the EIA of the Volta Grande Project with the Environment and Sustainability Agency (in Portuguese *Secretaria de Meio Ambiente e Sustentabilidade* [SEMAS]). The consultation letter states that "in order to provide SEMAS's environmental analysts with minimum information regarding the project and the area where it is slated to be implemented, Verena carried out a technical site visit to contextualize the location. Therefore, it proposes a TOR adapted to specific conditions of the localization of the project."[54] Thus, the consultation letter aims to provide basic information on the environmental baseline of the area to SEMAS and requests its approval on the draft TOR.

Therefore, it is possible to infer that an environmental feasibility study was carried out, although it was not attached to the consultation letter, as it is not a legal requirement. SEMAS issued the TOR on November 11, 2010.[55] There is no evidence of public participation in its development.[56] Additionally, the Belo Monte Dam was awarded its preconstruction permit in 2010. Considering the connection of both projects, which will have serious effects on the Volta Grande, synergetic impacts should have been assessed in both EIAs as to whether actions were well-planned and coordinated.

On February 2, 2012, Belo Sun filed the EIS and its nontechnical summary report in conjunction with a request for a preconstruction permit of the project.[57] The EIS had eight volumes. In 2013, the State Environmental Council (in Portuguese *Conselho Estadual do Meio Ambiente* [COEMA]) voted in favor of issuing the preconstruction permit for the Volta Grande Project.[58] However, the opinion of the council was not unanimous. The state *Ministério Público*, who was also a member of the COEMA, voted against it. Furthermore, the *Ministério Público* wrote an official opinion stating its legal reasons for rejecting the preconstruction permit authorizing the environmental viability of the Volta Grande Project.

The opinion alleged irregularities in the environmental permitting procedure of the Volta Grande and the *Ministério Público* pointed out some other issues. First, the TOR was not attached to the Environmental Permitting Procedure. According to the *Ministério Público* this breaches CONAMA No. 237/97; it also makes comparison of the EIS impossible with its guidelines (TOR) previously issued by the environmental agency. When the *Ministério Público* finally analyzed the TOR (after an official request was made to SEMAS, the TOR was sent to the state *Ministério Público*), it stated that the EIS was not in compliance.[59] Some requirements stated in the TOR, were not met, such as authorizations of the Indigenous Affairs Agency (in Portuguese *Fundação Nacional do índio* or FUNAI), license from the Ministry of the Army, and authorization of the National Institute of Historic and Artistic heritage (in Portuguese *Instituto do Patrimônio Histórico e Artístico Nacional* or IPHAN).[60] Furthermore, the project was planned on public lands (federal public forests) and agrarian land reform areas, which required the participation of the Land Reform Area Agency and the Brazilian National Forest Service.[61]

Despite these matters, and also due to the fact that legal actions were filed against the state of Pará and Belo Sun, the majority of COEMA's members approved the preconstruction permit of the project. The preconstruction permit was issued on February 20, 2014; it has thirty-nine conditionalities. Some of these conditions are actions that should have been done before issuance of the permit, especially considering that the objective of the pre-construction permit is to certify the environmental viability of the project in that specific area. Thus, conditions include requiring that Belo Sun consult the Land Reform Agency (in Portuguese *Instituto Nacional de Colonização e Reforma Agrária* or INCRA) and carry out studies of impact of the project on indigenous communities.[62] These measures directly affect the social and environmental viability of the project. Therefore, they should not be conditions on the permit.

The installation permit was issued on February 2017. Two subsequent injunctions suspended this permit. However, these decisions were reversed.[63] Again, in September 2018, the project was stalled due to a decision from the Federal Court determining that the entire permitting process should be transferred to IBAMA.[64] The project has faced criticisms from local miners and communities, nongovernmental organizations, and the *Ministério Público*. SEMAS then organized a meeting with local communities after the issuance of the preconstruction permit.[65] While the meeting was not a legal requirement, since public hearings should happen before a permit is issued, the objective was to explain the project to communities and minimize contrary reactions.

EIA OF THE VOLTA GRANDE PROJECT

The Volta Grande Project is marked by controversies and legal actions. Several issues, including deficient public participation within its EIA procedure, were questioned by the state and federal *Ministério Público*. Additionally, civil society expressed its concerns at the implementation of the activity at the Volta Grande area.[66]

From 2013 to 2014, four legal actions were filed at the Federal Court against the state of Pará and Belo Sun.[67] The main arguments were related to deficiencies pointed in the EIA of the project, such as

- absence of a comprehensive assessment of the synergic and cumulative impacts of both projects: Volta Grande and the Belo Monte Dam;[68]
- non-assessment of impacts of the Volta Grande Project on indigenous communities who live in the surrounding areas as well as lack of their free, prior and informed consent, according to the International Labor Convention No. 169;[69]

- incompetence of the state environmental agency to conduct the environmental permitting procedure;[70]
- Belo Sun's acquisition of land in the area of the project, and prohibition of public access into those lands.[71]

After 2014, other legal actions were filed against the project and up to 2019 there are in total six lawsuits against the Volta Grande Project.[72]

The lack of effective and meaningful public participation reflects both cause and effect of many problems pointed out in the EIA of the Volta Grande Project. In other words, an effective public participation would assist the control of quality of the EIA. For instance, matters such as absence of a comprehensive cumulative impacts study could be avoided by having full consultation with communities and all interested agencies. In this vein, the involvement of a small community in Mckees Mill, in Canada, at an early stage of a plan to take down the Highway 134 bridge over the Bouctouche River, identified the necessity to assess cumulative impacts of this activity within the EIA because the river had been suffering heavy siltation, shallow waters, reduced size of its channel, and loss of its significant part of its waterway eelgrass. Residents believed these effects were caused by the Highway 134 bridge and its causeway.[73]

In this case, community residents mobilized themselves and provided scientific fact (some residents had previous knowledge in chemistry) as well as their experience and observation of the river along the years (some residents lived on the margin of the river their entire lives). Community members formed an association to discuss and contribute to the planning of the region, by carrying out surveys asking residents their views and informing them on the effects the Highway 134 bridge as well as whether its causeway had been causing impacts on the water flow of the river. Alternatives to taking down the bridge and using other bridges or replacing the Highway 134 bridge were discussed with community members.[74] This case demonstrates that early community engagement at EIA processes benefits not only themselves but also governments' plans.

For the purpose of this book, only issues related to public participation are discussed in the next sections.

Lack of Participation from Important Stakeholders

Official documents demonstrates that the following stakeholders were not consulted, and therefore did not participate from the very beginning of the EIA procedure:[75] FUNAI;[76] the health agency (in Portuguese *Fundação Nacional de Saúde* [FUNASA]);[77] INCRA; Ministry of Agrarian Development (in Portuguese *Ministério de Desenvolvimento Agrário* or MDS); the Norte

Energia Consortium responsible for the installation and operation of the Belo Monte Dam; and IBAMA.[78] A brief explanation on the importance of participation from each of these stakeholders in the EIA of Volta Grande, as well as the reasons why it did not happen, is provided.

FUNAI and Indigenous Communities

As previously mentioned, Volta Grande is to be located in the center of three indigenous communities (Paquiçamba, Arara da Volta Grande, Ituna/Itata). Furthermore, some indigenous individuals, although they no longer live with their communities, still live in the region.[79]

Despite the existence of these indigenous lands, FUNAI was not consulted on the EIA, before its development. Nor was FUNAI invited to attend the public hearing to discuss the project. The TOR for the EIA was issued in 2010; however, only in 2012, following official requests from the *Ministério Público*, was FUNAI consulted. FUNAI then issued a specific TOR for the realization of assessment of impacts of the mining activities on the indigenous communities. However, it should be noted that, when FUNAI intervened in the procedure, the EIS was already completed, and one public hearing had already happened.[80]

Whether FUNAI should have been consulted in this case remains contentious. Interministerial Ordinance No. 419/11 (which was revoked in 2015 by Interministerial Ordinance No. 60/15) set geographical limits to determine whether indigenous communities would be affected by certain activities, therefore whether FUNAI should participate in the EIA or not.[81] To understand the problem better, which involves the participation of FUNAI at the beginning of the EIA of the Volta Grande Project, some background is necessary.

CONAMA Resolution No. 001/86 requires that the EIS define the geographical area directly or indirectly affected by the impacts of a project—the project's area of influence.[82] In the case of the Volta Grande Project, the EIS defined as directly affected area the extension of territory where the physical structure of the project would be implemented, such as the gold mining pit, the waste dump, and offices. These areas correspond to Ressaca village, Galo, and Ouro Verde gold-diggings.[83] The area of direct influence was also defined as areas that might put pressure on infrastructure and public services, such as traffic, education, health care, and public safety. These correspond to villages and artisanal gold mining areas.

One of the main sources of discussion in this case revolves around the area of indirect influence determined by the project EIS. The Paquiçamba and Arara da Volta Grande indigenous lands are included in the area of indirect influence of the project because, according to the EIS, the project will not interfere in the indigenous lands itself. The EIS also states that it will affect the indigenous communities only when they interact with local communities outside their land.[84]

As mentioned in chapter 5, in Brazilian legislation, there is no methodology to establish areas of influence and impact. In this regard, Technical Note No. 39—a study written by the *Ministério Público*—also points out how commonly EIA developers manipulate the area of influence to avoid including certain areas, such as indigenous lands:[85]

> [In]digenous lands have been excluded of areas of influence, based on the physical and direct impact. The indigenous lands were only considered when the EIAs concluded that the project would impact these lands directly (cause and effect), for example indigenous lands which are flooded by hydroelectric power plants. It was verified that, EIAs does not analyze, effectively, the cultural importance and the relationship between the areas which will be impacted by project and the indigenous communities, even when they are affected indirectly.

Aside from the area of influence, when indigenous lands are located in the surroundings of the planned project, Interministerial Ordinance (IO) No. 419/11 (replaced by IO No. 60/2015) must be observed. According to the ordinance, projects that will interfere in indigenous lands must have the participation of FUNAI.[86] The interference in indigenous lands happens when the project is located inside indigenous lands or it would cause social and environmental damage to their lands, respecting the distance of 10 kilometers.[87] In other words, projects located above the limit of 10 kilometers, according to Interministerial Ordinance No. 419/2011, do not need to consult FUNAI nor carry out studies of impacts on indigenous communities.

The problem is that Interministerial Ordinance No. 419/11 does not clarify how to measure the 10 kilometers. By reading the regulation, it seems it should simply be measured from the indigenous lands to the project. However, the project itself has an area of direct and indirect influence of impacts. Therefore, should the area of influence be considered in the calculation of the referred ordinance? Or should only the physical installation of the project be considered? There is no answer either in the Interministerial Ordinance or in any Brazilian legislation.

Thus, in the Volta Grande Project, SEMAS took the physical area of mineral exploration conceded to Belo Sun by DNPM as the point of measurement.[88] Therefore, the distance of the project to the closest indigenous lands, the Paquiçamba would be 12 kilometers. However, as pointed out by Instituto Socioambiental, the Paquiçamba community already had authorization to expand their land.[89] According to SEMAS, "[t]he distance between the project (considering the area demarcated by DNPM) and the Paquiçamba, already considering the expansion of this land, is still above 10 km, more specifically it is 10.7 km. Therefore, SEMAS' team assert that there is not any doubt regarding the compliance of the legal limits required by the Interministerial Ordinance No. 419/11."[90] Notably, the understanding on considering the physical area demarcated by the National Department of Mineral Production diverges from the *Ministério Público.*

In its written vote, the *Ministério Público* pointed out that the area of influence of the biotic environment should be considered by measuring the 10 kilometers stated in Interministerial Ordinance No. 419/11.[91] Therefore, the impacts of projects would be considered and not just its physical installation. In this case, the distance would be less than 10 kilometers. Thus, it would be necessary to consult FUNAI and the indigenous communities by having their free, prior, and informed consent to authorize the project.

Furthermore, due to lack of guidance, there are several interpretations of the distance of the closest indigenous lands to the project. Analyzing the environmental permitting procedure, the following different measurements exist: the distance from Paquiçamba land to the project is 9.5 kilometers (according to Socioambiental Insittute), 11.6 kilometers (according to SEMAS), and 9.2 kilometers (according to state *Ministério Público*).

The Federal Court decided that assessment of impacts on indigenous communities were essential components and determined the suspension of the environmental permitting procedure until the completion of these studies. The court stated that the geographical limits of Interministerial Ordinance No. 419/11 are not rigid and might be altered in certain circumstances, according to Article 2, paragraph 3 of the ordinance.[92] However, this decision was reversed by the court of appeal, which determined the obligation of developing the assessment of impacts on indigenous peoples after the issuance of the preconstruction permit.[93] Therefore, despite the fact that the objective of the preconstruction permit is to ensure the environmental viability of the project, the court of appeal's decision to authorize assessment of impacts on indigenous communities,after issuance of the permit, is an evident contradiction with the goal of the permit. According to SEMAS, "observing the precautionary principle the assessment of impacts on indigenous communities must be carried out after the issuance of the preconstruction permit."[94]

In summary, the area of influence of impacts, which is normally divided into direct and indirect areas of influence, aims to demarcate the area of study and where mitigation and compensation measures should apply. However, instead of shedding light on when indigenous communities should be consulted, the former Interministerial Ordinance No. 419/11 brought more confusion because it considered distance in kilometers and not the impacts per se, while the area of influence considers the impacts resulted from the project. It seems that Interministerial Ordinance No. 419/11 guarantees less protection to indigenous lands than to other communities (nonindigenous communities).

The absence of FUNAI at the commencement of the EIA had consequences throughout the procedure and also revealed the contradictions in Brazilian regulations regarding protection and consultation of indigenous communities; the absence of parameters to establish areas of influence of impacts; the importance of stakeholders' participation (including participation of governmental agencies) during the TOR—often minimized in practice, including by the judiciary; and that the system of the three stages of environmental permitting procedure, adopted in Brazil, is constantly being breached by the common practice of postponing conditions that should have been fulfilled from the previous permit when approving the next permit.

FUNASA

According to CONAMA Resolution No. 286, of August 30, 2001, projects planned to be located in areas endemic for malaria must involve FUNASA, which will delimit the guidance of studies that should be developed as well as measures to control malaria by the project proponent.[95] Such commandment was corroborated by Interministerial Ordinance No. 419/11, which states that FUNASA must participate in the EIA, by issuing a specific TOR, containing topics related to health and programs capable of preventing malaria within workers. The state *Ministério Público* written vote sets forth that FUNASA was not consulted in the TOR of the Volta Grande's EIA.[96]

INCRA and MDS

The area where Belo Sun plans to implement the Volta Grande Project is public land. Thus, INCRA and MDS should have been consulted regarding the area of the project.[97] In the public hearing carried out in September 2012, Belo Sun stated that INCRA was not consulted because the agency was on strike.[98] Yet, INCRA was not even invited to the public hearing to discuss the EIS of the project; INCRA was only consulted in November of 2012—after the first public hearing. Notably, Interministerial Ordinance No. 419/11, which sets the norms on consultation of agencies at the commencement of the EIA, did not include any provisions on consultation with INCRA and MDS, whenever the project is located in agrarian reform area.[99]

Norte Energia and IBAMA

Norte Energia did not acknowledge the EIA of Belo Sun from the outset. However, its participation was important because the Volta Grande is in the area of direct influence of Belo Monte; therefore, Norte Energia is responsible for developing several social and environmental programs in this area.

Inclusion of Norte Energia was actually even more fundamental given its mandate for Belo Monte and the impacts of Volta Grande. As a conditionality of the Belo Monte permit, Norte Energia refurbished the public school at the Ressaca village as well as started implementation of the sanitary sewage in the village. However, the EIS of the Volta Grande Project discusses the necessity of resettling the Ressaca village. Thus, while Norte Energia was improving conditions for the lives in the area, the state of Pará was looking at the possibility of resettling the entire village. This example demonstrates the lack of territorial and development planning in the scope of EIA procedure of projects located in the same region.

Later, Norte Energia presented an extensive letter to SEMAS requiring the suspension of the Permitting Procedure of the Volta Grande Project due to the incompatibility of both projects in the Volta Grande.[100] According to Norte Energia, the Belo Monte Installation Permit has eight conditionalities specifically related to the Volta Grande area. These conditionalities are related to monitoring the portion of the river where the water flow will substantially decrease (the Volta Grande). Norte Energia stated that "the fragility resulted from the reduction of water flow in the Volta Grande shall be monitored by Belo Monte's entrepreneur [Norte Energia], which is incompatible, at first sight, with a territorial concurrence of another project, authorized ignoring the conditionalities of the Belo Monte Dam."[101] Essentially, another project in the area, in addition to generating new impacts on the Volta Grande, will interfere with actions taken by Norte Energia. Moreover, a project implemented in the Volta Grande requires, among others, the consultation to indigenous communities and the coordination of the social and environmental programs in the Volta Grande as well as consultation to IBAMA on the restrictions of utilization and occupation of the Belo Monte area.[102]

As a result of this lack of coordination of agencies and the absence of participation of IBAMA, on November 2014 the Federal Court determined that IBAMA must participate in all phases of the permitting procedure of the Volta Grande Project.[103] However, this decision was reversed in the second instance; under Brazilian law, there is no possibility of carrying out environmental permitting procedure by two environmental agencies (in this case SEMAS and IBAMA). Therefore, SEMAS was determined competent to conduct the procedure and IBAMA was not included in any phase of the environmental permitting procedure. According to the high court, such measure would cause "excessive bureaucracy, increase costs, postpone decisions and create the possibility of conflicting decisions, when such environmental permitting procedure should have been conducted by only one federal entity [SEMAS] within the limits of its competence [. . .] and it does not guarantee more protection or benefit to the environment."[104]

The competence of the permitting procedure was discussed again at the Federal Court and, in September 2018, the court ordered that the permitting procedure of the Volta Grande Project be transferred to IBAMA, as the only agency competent to conduct the permitting procedure. The project was stalled, awaiting the court to authorize its continuation.[105]

The participation of IBAMA since the beginning of the EIA was of crucial importance, especially considering its responsibility to monitor the amount and quality of the Volta Grande for six years.[106] In this respect, Article 13 §1 of Federal Law No. 140/11 states that

> Projects and activities are subject to environmental permitting procedure or authorized, by one federative entity, in conformity with its attributions, according to this law. §1 Other federal entities interested in the permitting procedure may express their opinions to the responsible agency, in a non-binding manner, respecting deadlines and procedures of the environmental permitting.[107]

Therefore, a work in conjunction of state and federal environmental agencies would be legal and possible. Furthermore, it would benefit the environment due to the proximity of both projects and their impacts on a fragile region (the Volta Grande area). Therefore, coordination of actions in the Volta Grande would be important to plan the future of the region in a responsible and sustainable manner.

Access to Information and Communities' Participation in the EIA of the Volta Grande Project

Access to information is key to meaningful public participation because stakeholders can only participate and provide their input if previously informed. In essence, stakeholders should have access to the EIS prior to any public discussion of the project. It is also necessary that contents are written in language comprehensible to laypersons and that the communication of public hearings is accessible to interested stakeholders. This section demonstrates how the right of access to information is sometimes denied as an ordinary practice. Geographical limitations, illiteracy, and technical language often pose barriers to this right. Governments should pay close attention in order to guarantee that communities are fully informed during a consultation process.

The TOR of the Volta Grande Project's EIS

The Volta Grande Project demonstrates that the scoping of the EIA is seen as a separate or unimportant phase of the procedure. According to the state

Ministério Público, when it reviewed the EIS before its approval, the TOR was not attached to the Environmental Permitting Procedure, which made it difficult to compare and evaluate the EIS with its guidance.[108] SEMAS stated that the TOR was not attached to the EIS because it was requested by Verena, which was later acquired by Belo Sun. In fact, even though SEMAS gave this explanation, it is not clear why exactly the TOR was not attached to the EIS.

The TOR of the Volta Grande EIA was issued in 2010 and the EIS was filed in 2012. For the purpose of defining the competence of the environmental agency (whether state or federal agencies should conduct the process) the Federal Court decided that the environmental permitting procedure started when the EIS was filed, completely ignoring the scoping phase, which is a very important step of the EIA procedure.[109] The court stated that although Belo Sun filed a preliminary study of the area and requested the issuance of the TOR in October 2010, it was not possible to understand that requesting the issuance of the TOR was part of the environmental permitting procedure. According to this understanding, the environmental permitting procedure starts only when the environmental agency provides its technical opinion, which happens when the company files the EIS.[110] The elaboration of the TOR was not open to public input nor was it considered in this case as the first step of the EIA procedure. The EIS was totally separated from its origin—the TOR.

Access to the EIS and Its Nontechnical Summary

The EIS and its nontechnical summary are public documents and its prior consultation is very important to provide a basis for participation in the public hearing. In the Volta Grande Project, the EIS and its nontechnical summary were available for consultation in the SEMAS, which is located in Belém, the capital of Pará State—826 kilometers away from SJP. Communities in SJP, who would be the most interested in consulting the EIS and its nontechnical summary, could not do so, considering their physical inaccessibility.

CONAMA Resolution No. 001/86 requires that the nontechnical summary should be made available to the public.[111] Respecting industrial privacy, copies of the nontechnical summary should be available at the documentation centers or environmental agencies' library.[112] Therefore, SEMAS did not breach CONAMA Resolution No. 001/86 when it made the EIS and its nontechnical summary available in its library, although it is 826 kilometers away from the affected communities. However, CONAMA Resolution No. 001/86 obviously needs to be clearer about the disclosure of EIS, meaning that it should be made public where affected communities are able to consult them rather than restricting to a certain place (the library of the environmental agency).

In this case, and in many others, the environmental agency should have provided the EIS and nontechnical summary copies in the community center of the municipality where the impacts will be most felt. Although CONAMA Resolution No. 001/86 states that the EIS and its nontechnical summary should be available at documentation centers and the environmental agency's library, there is not a prohibition to make them available at the municipality. Furthermore, the Federal Constitution states that EIS should be made public. Thus, a truly public document is a document which can be accessed by all, including the most interested parties. The realization of the public hearing scheduled for September 12, 2012, was published in the Official Gazette forty-seven days prior to the hearing.[113]

The first public hearing was held in the municipal center of Senador José Porfírio (it is also the urban core of SJP); SEMAS sent several invitations to public authorities to participate in this public hearing.[114] However, all invitees were located in Belém. Community centers from SJP were not invited, and neither were FUNAI, FUNASA, INCRA, nor the municipality's authorities. The public hearing was attended by 237 people; but the *Ministério Público* pointed out that the number of attendees did not reflect the number of potentially affected people in the area. Below, some reflections on the dialogues that occurred at the public hearings are made.

Public Hearing of September 12, 2012

Several questions were raised during the hearing, including some important dialogues. For instance, a representative of INCRA, who was not formally invited, but voluntarily attended, questioned whether the project is planned to be located in area of agrarian reform and whether INCRA was consulted. Belo Sun's representative responded that they tried to consult INCRA, but the agency was on strike. However, Belo Sun intended to contact INCRA.[115]

A former mayor of SJP stated that "R$ 2.7 million were paid in taxes and so far they have not seen any improvement as a result of the money received."[116] Most importantly, he asked about the use of explosives on the rocks to extract gold and whether these explosions would affect the Belo Monte Dam. He asked the company to meet and discuss with communities these issues as well as what cumulative and synergic impacts studies should be carried out. Belo Sun's response was that the company is carrying out studies in relation to the explosions of rocks and the dam. However, Belo Sun's representative assured that there would be no interference of explosions on the dam because there is a "massive sealant."[117] Notably, technical language was used to explain the nonoccurrence of impacts and no further explanation was provided.

Those who live close to a massive dam and a mining project would naturally fear the burst of the dam, which would flood their houses. Therefore, appropriate language to explain the impacts or its nonoccurrence is necessary to make the most affected people aware of their future in the area. The fear that communities should also be considered and managed in the consultation process of EIAs. Another point is that the company is asking for authorization of a preconstruction permit which authorizes the environmental feasibility of the project; however, it is still carrying out studies to ensure that there will be no impact on the dam. As previously mentioned, there is a contradiction in the practice of a three permits system. How would the environmental agency ensure the environmental feasibility if such an important study was still ongoing? Although SEMAS explained that there is a difference between risks and impacts and that the referred study would be a study on the risk of influence of both projects, it does generate insecurity among communities and diminishes the purpose of a preconstruction permit.

In relation to the language utilized, the mayor of SJP asked that next time the company met with affected communities they utilize more accessible language because the technical language they were using would not be understood by many people. The mayor also said that many questions were not addressed in the hearing because there was not enough time for the discussions. Therefore, more time was necessary to answer the questions of the people, according to him.[118] Belo Sun answered that they would meet with the potentially affected communities to discuss and answer questions.

The prosecutor Thais Santi stated that she did not recognize that meeting as an official public hearing because it was held distant from affected communities and they could not participate due to the difficulties in accessing the place.[119] This brief reproduction of dialogues during the September 2012 hearing tries to demonstrate that the communication was not effective due to many reasons: the place chosen was far from affected communities, therefore did not guarantee their participation; the language utilized was not appropriate to laypersons; the time was not enough to discuss important issues, which should have been discussed before this stage, where the EIS was already written. It is possible to infer that all the discussions of an important project were left to the public hearing where many questions were not answered properly. Although the company committed to meet with affected communities, previous meetings would have been very effective in this case and not meeting after an official public hearing.

Public Hearing of January 10, 2013

After the September 2012 public hearing, Belo Sun filed technical notes clarifying doubts raised during the hearing. SEMAS scheduled another public hearing for December 6, 2012, in Ressaca village. Belo Sun published the occurrence of the public hearing in official gazettes and the local newspaper.

The EIS and its nontechnical summary were available at SEMAS library as usual. In November 2012 the state public defenders wrote an official recommendation to SEMAS requesting the following: the technical notes clarifying doubts about the EIS should have been made available for public consultation in accessible language to laypersons; the postponement of the public hearing, based on the fact that the EIS and its nontechnical summary were not available to Ressaca village residents' consultation and pointing out that public defenders only had access to the EIS because they flew to Belém (less than thirty days before the hearing). The EIS, nontechnical summary, and technical notes should be available at Ressaca village providing time (at least thirty days) prior to the public hearing to consultation. Public defenders also stated that many individuals were not aware of the date of the public hearing in Ressaca village; therefore, they kept asking at the public defenders' office.[120]

In response, SEMAS postponed the public hearing to January 2013. It determined that Belo Sun would make the EIS, the nontechnical summary, and technical note available to communities at Ressaca village and at other locations in the municipality. However, in relation to technical notes SEMAS stated that they were just clarifications; therefore, it was not necessary to rewrite them in accessible language. According to SEMAS the content of the technical note was already in the EIS and its nontechnical summary. In addition, they were presented by Belo Sun in order to clarify doubts of the *Ministério Público*. Therefore, the mentioned institutions and others would receive the technical notes.

Prior to the public hearing in January 2013, SEMAS requested Belo Sun to advertise the hearing on the radio. The company advertised on local radio twenty-four days prior to the hearing (in total fifty-seven radio advertisements). Belo Sun also put advertising slots around Ressaca village, announcing where the EIS and its nontechnical summary report could be found in order to be consulted prior to the hearing. These documents and the EIA technical note were available at the Mining Cooperative in Ressaca village, at the Public Elementary school at Ressaca village, at the City Hall of SJP, and at the SEMAS regional office in Altamira.[121] After these measures, the public hearing occurred. There were 441 people in attendance—more than 53.74 percent attended the first hearing, even though it took place in the rural core of SJP.

During the public hearing, Belo Sun stated that they carried out twenty previous meetings in Belém, SJP, Galo, Ressaca, Itata and Fazenda villages, and in Altamira. Approximately 700 people attended these meetings and around ninety questions were answered, according to the company.[122] However, in interviews with local communities, public defenders, and members of civil society they all said that these meetings were not useful in explaining the impacts of the project.[123] Additionally, they said that this was a tentative of cooptation of the company. Obviously, all these interviewed stakeholders were against the project for their own rationale. However, SEMAS did not participate in these previous meetings, which means only the company and probably the consultancy that carried out the EIA and wrote the EIS met with communities. Therefore, chances to advertise the projects benefits in detriment to real negative impacts are great in such situations. A neutral facilitator would have an important role in meetings with communities.

In sum, the EIA procedure of the Volta Grande Project demonstrates serious gaps in Brazilian regulation concerning access to information and public participation. Additionally, in general, there was no effort to act differently or interpret regulations in a manner that provides access to all interested parties. The fact that communities constantly sought public defenders to enquire about the date of public hearing demonstrates how weak the publication in official gazettes and local newspapers is in the role of advertising the hearings. It is also important to highlight that 30 percent of SJP population above fifteen years of age are illiterate.

Good practice was in place when Belo Sun advertised on the radio and spread banners in the municipality to make communities aware of the availability of the EIS and its nontechnical summary.[124] Unfortunately, this happened after significant disturbance in the procedure. Belo Sun's initial plans were to start installation of the project at the beginning of 2015.[125] As of the time of writing the installation permit is suspended. Considerable delay in the time line of the project can partially be attributed to legal actions against the project. Furthermore, in May 2013, the price of Belo Sun's shares dropped by up to 40 percent, representing a price fall of over 60 percent a year. According to Jacobi, the market did not react well to the news of problems involving the project such as the proximity and potential cumulative impacts with Belo Monte Dam, impacts on indigenous communities, and conflicts with artisanal miners. So many issues could make the project much more expensive than it was predicted by Belo Sun in its economic feasibility report published on its website.[126]

Notably, if it was not for the recommendation of the public defenders, the environmental studies would not be available at Ressaca village, the public hearing would not be postponed, and radio advertisements and banners would not be posted. It would have been another public hearing where communities did not consult the EIS or worse: they were unaware of its existence, prior to the hearing. Clearer regulations are necessary along with a mechanism capable of monitoring whether access to information is guaranteed from the beginning of the procedure. In this regard, the role of independent committees, discussed in chapter 8, would be of great importance in avoiding delays and making the EIA procedure more accessible and participation more effective.

CONCLUSION

The region of the Volta Grande Project is unique for many reasons. It is rich in natural and cultural resources, where different social groups live and interact, such as indigenous and riverbank communities, artisanal miners, and villagers. The area is economically poor and has a high occurrence of malaria. Aside from these characteristics, it is also public land within the area of influence of the megaproject Belo Monte Dam. In sum, this is a region where different stakeholders should interact and participate in the process of decision-making regarding any development.

The lack of coordination of important stakeholders demonstrates serious shortcomings of the Brazilian EIA Procedure. As demonstrated in this chapter, the existing regulations are not clear enough to provide guidance on the identification of stakeholders. Furthermore, in the case of indigenous communities it seems that the only regulation available creates more doubts than provides guidance on participation of FUNAI and the indigenous peoples who live in the surrounding areas of development projects. Some measures, suggested in upcoming chapters, would assist in the identification of relevant actors involved in the implementation of projects.

Access to information and public participation are naturally denied when EIS and its nontechnical summary are made available in distant locations, intending just to comply with the regulation. Considering this example, it is possible to infer that there is no real intention to inform communities prior to their participation in public hearings. Compliance with commandments in CONAMA Resolution No. 001/86 are carried out as a "tick-the-box approach" and efforts to make EIS and its nontechnical summary available for most affected communities are only made after the intervention of legal institutions (in this case, the public defender). Moreover, other means of communication—apart from publication in official gazettes and local newspapers—are necessary to disseminate the information. Access to information is also denied when technical language is utilized to explain the project and its impacts to affected communities.

Trust in the environmental permitting system is also an issue in Brazil. Conditionalities are used as means to move permitting procedures forward. Considering that these conditions are not often fulfilled and, essentially, are repeated on subsequent permits—as happened in the Belo Monte case and in many other projects in Brazil. The fact that conditionalities are often not "true conditions" but legal "maneuvers" to push permitting procedures forward despite serious concerns breaks people's trust, giving them a sense that in the end, even the most relevant issue can become a conditionality, which will be repeated in other permits. However, they do not impede development of the project and ultimately its approval. For instance, in the

Volta Grande permit, some conditionalities were actions that should have been done previously.

In this case, communities are divided in relation to the installation of the project in a region where they have been living for many years. The main objective is not to question whether the state of Pará should allow exploration of gold in this region; however, the goal is to demonstrate the lack of effective dialogue and meaningful community participation in a situation where their voices could have had a heavier weight in the final decision of the project's authorization. The following chapters will explore Brazilian EIA legislation and the opportunities stakeholders have to participate in this procedure, in particular the ways to improve public participation in a worldwide comparative analysis.

NOTES

1. Belo Sun Mining, Overview, last accessed November 23, 2019, https://www.belosun.com/about_us/overview/.
2. See Secretaria Estadual de Meio Ambiente e Sustentabilidade, *Processo de Licenciamento Ambiental No. 2012/0000005028* (Technical Opinion, Environmental Permitting Procedure) (Parecer Técnico 22.520) (on file with author).
3. "Cidades, Pará, Senador José Porfírio" (Cities, Pará, Senador José Porfírio), Instituto Brasileiro de Geografia e Estatistica (IBGE), last accessed October 14, 2019, http://cidades.ibge.gov.br/xtras/temas.php?lang=&codmun=150780&idtema=16&search=para|senador-jose-porfirio|sintese-das-informacoes.
4. Ibid., GDP until 2016; U.S. dollars based on 2016 rate.
5. Brandt Meio Ambiente, *Estudo de Impacto Ambiental do Projeto Volta Grande* (Environmental Impact Statement of the Volta Grande Project) (2010), Item 5 (on file with author).
6. Ibid.; IBGE, "Cidades, Pará, Senador José Porfírio."
7. "Estudo de Impacto Ambiental do Projeto Volta Grande," Item 5.
8. Brandt Meio Ambiente, *Relatório de Impacto Ambiental: Projeto Volta Grande* (Environmental Impact Assessment Non-Technical Summary: Volta Grande Project) (2012).
9. Ibid.
10. Information gathered during field research trip from July 1–9, 2015.
11. Maria Celia Nunes Coelho et al., "Extrativismo do Ouro no Século XXI. Exemplos no Sudoeste da Amazônia Brasileira" (Gold Extrativism in the XXI Century. Examples at the Southeast of the Brazilian Amazon), *Anuario do Instituto de Geografia UFRJ* 39, no. 3 (2016): 5, http://www.anuario.igeo.ufrj.br/2016_3/2016_3_05_14.pdf.
12. "Relatório de Impacto Ambiental: Projeto Volta Grande."
13. A. O. Sawakuchi et al., "The Volta Grande do Xingu: Reconstruction of Past Environments and Forecasting of Future Scenarios of a Unique Amazonian Fluvial Landscape," *Scientific Drilling* 20 (December 2015): 21.

14. "Estudo de Impacto Ambiental do Projeto Volta Grande," Item 5.

15. Leme Engenharia,"Relatório de Impacto Ambiental—RIMA da Hidrelétrica de Belo Monte" (Environmental Impact Statement Non-Technical Summary of Belo Monte Dam), http://restrito.norteenergiasa.com.br/site/wp-content/uploads/2011/04/NE.Rima_.pdf.

16. Portaria do ICMBIO No. 9, de 23 de Janeiro de 2007, Diário Oficial da União (D.O.U.), 24.01.2007 (Braz.). See also Ministério do Meio Ambiente, *Áreas Prioritárias para a Conservação Uso Sustentável e Repartição de Benefícios da Biodiversidade Brasileira (AM179)* (Priority Areas to the Conservation and Sustainable Use and Sharing of Benefits of the Brazilian Biodiversity AM179), January 23, 2007, http://www.mma.gov.br/estruturas/chm/_arquivos/biodiversidade31.pdf.

17. "Estudo de Impacto Ambiental do Projeto Volta Grande," Item 5.

18. Ibid.

19. Ibid.

20. J.F., ACP N. 0001739-80.2014.4.01.3903, Subseção Judiciária de Altamira, 29.08.2014 (Braz.). Although there are several different measurements indicating the distance of the project to indigenous communities, the measurements pointed above are included in the EIS of the project.

21. "Belo Monte Usina Hidrelétrica," Agência Nacional de Energia Elétrica, last accessed November 27, 2019, http://www2.aneel.gov.br/aplicacoes/hotsite_beloMonte/index.cfm?p=7.

22. "Usina a Fio D'agua e Menor Area Alagada" (Hydro Power Plant and Less Flooded Area), Usina Hidreletrica Belo Monte, Norte Energia, last accessed November 23, 2019, https://www.norteenergiasa.com.br/pt-br/uhe-belo-monte/arranjo.

23. "Fact Sheet: The Belo Monte Dam," Amazon Watch, last accessed November 23, 2019, https://amazonwatch.org/assets/files/2011-august-belo-monte-dam-fact-sheet.pdf.

24. Ministério do Planejamento. "As 10 Maiores Obras do PAC" (The 10 Biggest Projects of the Accelerated Growth Program) July, 2013, http://pac.gov.br/i/b8.

25. See Indigenous Communities of the Xingu River Basin, Pará, Brazil, Precautionary Measures, No. PM 382/2011, Inter-Am. Comm'n H.R., http://www.cidh.oas.org/medidas/2011.port.htm

26. Katie Soltis, "Brazil Disregards Charges from the Inter-American Commission on Human Rights," *Council on Hemispheric Affairs*, June 9, 2011, http://www.coha.org/brazil-disregards-charges-from-the-inter-american-commission-on-human-rights/.

27. Indigenous Communities of the Xingu River Basin, Pará, Brazil, Precautionary Measures.

28. Ibid.

29. MPF, "MPF Recomenda ao Governo do Pará que Suspenda todos os Licenciamentos na Volta Grande do Xingu" (Ministerio Público Recommends that the Government of Pará Suspends all the Environmental Permitting Procedures at the Xingu Volta Grande), last accessed November 23, 2019, http://www.mpf.mp.br/pa/sala-de-imprensa/noticias-pa/mpf-recomenda-ao-governo-do-para-que-suspenda-todos-os-licenciamentos-na-volta-grande-do-xingu.

30. IBAMA, Technical Note sent to Norte Energia, July 6, 2015 (on file with author).

31. "Relatório de Impacto Ambiental—RIMA da Hidrelétrica de Belo Monte."

32. IBAMA, Technical Note sent to Norte Energia, November 2014 (on file with author).

33. Eve Bratman, "Passive Revolution in the Green Economy: Activism and the Belo Monte Dam," *International Environmental Agreements: Politics Law & Economics* 15, no. 1 (March 2015): 61.

34. "Justiça Suspende Licença de Belo Monte Atendendo a Pedido do MPF" (Court Suspends the Per of Belo Monte as Requested by Federal *Ministério Público*), *G1*, August 31, 2016 10:27 p.m., http://g1.globo.com/pa/para/noticia/2016/08/justica-suspende-licenca-de-belo-monte-atendendo-pedido-do-mpf.html.

35. Ibid.

36. Rodrigo Polito, "Justiça Cassa Liminar que Suspendia Licença de Operação de Belo Monte" (Court Rules Injunction that Suspended Belo Monte Operation Permit), *Valor Econômico*, September 16, 2016, 9:18 a.m., http://www.valor.com.br/empresas/4711981/justica-cassa-liminar-que-suspendia-licenca-de-operacao-de-belo-monte.

37. "Relatório de Impacto Ambiental—RIMA da Hidrelétrica de Belo Monte."

38. Ibid.

39. "Estudo de Impacto Ambiental do Projeto Volta Grande," 4, 5, 9.

40. Ibid., 4.

41. Ibid.

42. Ibid.

43. "Estudo de Impacto Ambiental do Projeto Volta Grande," 40.

44. Ibid., 2.

45. Lei No. 8.001, de 13 de Março de 1990, Diário Oficial da União (D.O.U.) de 14.03.1990 (Braz.).

46. Maria Amélia Rodrigues da Silva Enriquez, "Royalties da Mineração, Instrumento de Promoção do Desenvolvimento de Regiões Mineradoras da Amazônia Oriental" (Mining Royalties, Instruments to Promote Development in the Mining Regions of the Eastern Amazon), *Novos Cadernos NAEA, Belém* 1, no. 2 (1998), http://www.repositorio.ufpa.br:8080/jspui/bitstream/2011/3086/1/Artigo_RoyaltiesMineracaoInstrumento.pdf.

47. Ibid.

48. Ibid.

49. "Estudo de Impacto Ambiental do Projeto Volta Grande," 40, 41.

50. Ibid.

51. Instituto Socioambiental, *Análise do Licenciamento Ambiental do Projeto Volta Grande da Belo Sun Mineração* (Analysis of the Environmental Licensing of the Volta Grande Project owned by Belo Sun Mineração) (2013), https://www.socioambiental.org/sites/blog.socioambiental.org/files/blog/pdfs/2._manifestacao_belo_sun_isa.pdf.

52. Ibid.

53. Public Hearing held at Ressaca village in January 2013. See Volta Grande Environmental Permitting Procedure No. 5028/2012.

54. Ibid.

55. Ibid.

56. More recently SEMAS started posting the TOR of some projects for public consultation on its website. This was not a current practice at the time of the TOR for the Volta Grande Project.

57. Volta Grande Environmental Permitting Procedure No. 5028/2012.

58. COEMA is a deliberative and consultive council, which discusses environmental matters, such as approving the issuance of permits in the state of Pará. COEMA is composed of thirteen members who represent the government and the civil society. Lei Estadual No. 5610 de 20 de Novembro de 1990, Diário Oficial do Estado do Pará (D.O.E.PA) de 20.11.90 (Braz.).

59. Voto Escrito do Ministério Público sobre o Licenciamento Ambiental do Projeto Volta Grande (Written vote of *Ministério Público* on the Environmental Permitting Procedure of the Volta Grande Project) (on file with author).

60. Volta Grande Environmental Permitting Procedure No. 5028/2012.

61. Written vote of *Ministério Público.*

62. Ibid.

63. Eve Bratman and Cristiane Bena Dias, "Development Blind Spots and Environmental Impact Assessment: Tensions between Policy, Law and Practice in Brazil's Xingu River Basin," *Environmental Impact Assessment Review* 70 (May 2018).

64. TRF-1, No. 0001813-37.2014.4.01.390, Relator: _ 09.09.2018, http://www.mpf.mp.br/pa/sala-de-imprensa/documentos/2018/sentenca_justica_federal_licenciamento_ibama_projeto_volta_grande_mineracao_belo_sun_pa_setembro_2018.pdf.

65. "Mineradora Canadense Comprou Terras Irregulares na Amazônia" (Canadian Miner Bought Untitled Lands in the Amazon), *MAB Amazônia*, February 25, 2016 4:01 a.m., http://www.mabnacional.org.br/noticia/mineradora-canadense-comprou-terras-forma-irregular-no-xingu (last accessed November 6, 2016).

66. Instituto Socioambiental wrote a letter explaining its concerns as to whether the gold mining project was approved to operate in the Big Bend. The letter was handed over during the public hearing of the Volta Grande Project. See "Belo Sun," Instituto Socioambiental, last accessed January 14, 2017, https://www.socioambiental.org/pt-br/tags/belo-sun.

67. Justiça Federal 1 Região (Regional Federal Court 1st Region), last accessed February 10, 2015, http://portal.trf1.jus.br/sjpa/. Among four public civil actions filed at the Federal Court, only three of them are discussed in this manuscript.

68. J.F., ACP No. 0002505-70.2013.4.01.3903, Subseção Judiciária de Altamira, 11.11.2013 (Braz.).

69. Ibid.

70. J.F., ACP. No. 000183-37.2014.4.01.3903, Subseção Judiciária de Altamira, 29.08.2014 (Braz.).

71. INCRA Memorandum No. 0097/13 declares that the area bought by the company is public land; therefore, it is under INCRA's responsibility. Thus, the

land could not be acquired by the company. INCRA memo can be found at J.F., No. 0001739-80.2014.4.01.3903, Subseção Judiciária de Altamira, 29.08.2014 (Braz.).

72. Ministério Público Federal, *Processos Caso Belo Sun* (Lawsuits Against Belo Sun), last accessed November 24, https://docs.google.com/spreadsheets/d/1P4dXLzCpCPFNGlSarPFlkwDEygx3h4mgJmwVo_26sew/edit#gid=0.

73. Daniel Delong and Michael Fox, "Small Communities and Public Participation in the Cumulative Environmental Impact Assessment Process: The Case of Little Bouctouche River New Brunswick," *Journal New Brunswick Studies* I (2015): 13, http://w3.stu.ca/stu/sites/jnbs/en/issue_06-1.html.

74. Ibid.

75. Written Vote of *Ministério Público* on the Environmental Permitting Procedure of the Volta Grande Project; J.F., ACP No. 0002505-70.2013.4.01.3903.

76. J.F., ACP No. 0002505-70.2013.4.01.3903.

77. Written Vote of *Ministério Público.*

78. Ibid.

79. Ibid.

80. Ibid.

81. Portaria Interministerial No. 419, de 26 de Outubro de 2011, Diário Oficial da União (D.O.U.) de 28.10.2011 (Braz.). Chapter 6 discusses the Interministerial Ordinance No. 419/2011 (Braz.). Although the Interministerial Ordinance (IO) No. 419 was revoked by IO No. 60/15, this chapter refers to the first one because it was in force during the time of the events discussed.

82. Resolução CONAMA No. 001, de 23 de janeiro de 1986, Diário Oficial da União (D.O.U.) de 17.02.1986 (Braz.).

83. "Estudo de Impacto Ambiental do Projeto Volta Grande," Item 5, 7.2.1.

84. Ibid., 7.2.3.

85. Procuradoria Geral da República, *A Área de Influência no EIA: Análise e Propostas* (The Area of Influence in Environmental Impact Assessments. Analysis and Proposals) (4CCR PGR, Nota Técnica 039/2007, October 17, 2007), http://www.mpf.mp.br/atuacao-tematica/ccr4/dados-da-atuacao/grupos-de-trabalho/gt-licenciamento/documentos-diversos/informa-tecnicas/NT%20039-07_Encaminha_trabalho_area_de_influencia.pdf/view.

86. Portaria Interministerial No. 419/2011, art. 6.

87. Ibid., Annex II.

88. "Estudo de Impacto Ambiental do Projeto Volta Grande," Item 7.

89. Analysis of the Environmental Licensing of the Volta Grande Project owned by Belo Sun Mineração.

90. Volta Grande Environmental Permitting Procedure No. 5028/2012.

91. Written Vote of *Ministério Público.*

92. Portaria Interministerial No. 419/2011, art. 2, 3. "In exceptional cases, since it is appropriately justified and also considering the specificities of the activity and the local peculiarities, the limits of Annex II can be altered. However it is necessary to have agreement among IBAMA, the agency involved in the procedure and the project proponent."

93. TRF-1, A.I. No.11749-24.2015.4.010000/Pa Pará, Relator: Des. Jirair Aram Megueriam, 24.04.2015 (Braz.).

94. Volta Grande Environmental Permitting Procedure No. 5028/2012.

95. Resolução CONAMA No. 286, de 30 de Agosto de 2001, Diário Oficial da União (D.O.U.) de 17.12.2001 (Braz.).

96. Portaria Interministerial No. 419/2011.

97. J.F. No. 0001739-80.2014.4.01.3903.

98. Volta Grande Environmental Permitting Procedure No. 5028/2012, Transcription of the Public Hearing.

99. Interministerial Ordinance No. 419/11 states that INCRA should be communicated on projects planned to be implemented in Quilombola lands. It was revoked by Interministerial Ordinance No. 60/15, which also did not require consultation to INCRA whenever a project would be located in land reform area. Portaria Interministerial No. 60, de 24 de Março de 2015, Diário Oficial da União (D.O.U.) de 25.03.2015 (Braz.).

100. Volta Grande Environmental Permitting Procedure No. 5028/2012.

101. Ibid.

102. J.F., ACP No. 183-372014.4.01.3903, Subseção Judiciária de Altamira, 29.08.2014 (Braz.).

103. Ibid.

104. TRF-1, A.I. No.11749-24.2015.4.010000/Pa Pará.

105. TRF-1, No. 0001813-37.2014.4.01.390, Relator: _ 09.09.2018, http://www.mpf.mp.br/pa/sala-de-imprensa/documentos/2018/sentenca_justica_federal_licenciamento_ibama_projeto_volta_grande_mineracao_belo_sun_pa_setembro_2018.pdf.

106. The six years of monitoring is a condition of the Belo Monte Dam.

107. Lei Complementar No. 140/11, de 8 de Dezembro de 2011, art.13 §1, Diário Oficial da União (D.O.U.) de 9.12.2011 retificado em 12.12.2011 (Braz.).

108. Written Vote of *Ministério Público.*

109. J.F., No. 0001813-37.2014.4.01.3903, Subseção Judiciária de Altamira, 29.08.2014 (Braz.).

110. Ibid.

111. Resolução CONAMA No. 001, de 23 de Janeiro de 1986, art. 11, Diário Oficial da União (D.O.U.) de 17.02.1986 (Braz.).

112. Ibid.

113. Volta Grande Environmental Permitting Procedure No. 5028/2012.

114. Ibid.

115. Ibid.

116. Ibid.

117. Ibid.

118. Ibid.

119. Ibid.

120. Volta Grande Environmental Permitting Procedure No. 5028/2012, Vol. 3.

121. Ibid.

122. Volta Grande Environmental Permitting Procedure No. 5028/2012, Vol. 4.

123. Interviews with citizens and public officials involved with the environmental permitting of the Volta Grande Project, in Altamira, Senador José Porfírio, and Belém, Pará State, Brazil, July 2–9, 2015.

124. Volta Grande Environmental Permitting Procedure No. 5028/2012.

125. Ibid.

126. Pedro Jacobi, "Ações da Belo Sun Fecham em 70c Após Mercado Reagir Mal ao Estudo de Viabilidade Econômica" (Shares of Belo Sun are at 70c After Market Had a Bad Reaction to the Economic Feasibility Study), *O Portal do Geólogo*, May 7, 2013, 6:46 p.m., http://www.geologo.com.br/MAINLINK2013.ASP?VAIPARA=Acoes%C2%A0da%20Belo%20Sun%20fecham%20em%2070c%20apos%20mercado%20reagir%20mal%20ao%20estudo%20de%20viabilidade%20economica.

Part III

EIA STEPS

Chapter 4

Planning the Project and Engaging with the EIA Procedure

PLANNING THE PROJECT

Development may result from a market-driven necessity or willingness of private or public companies.[1] Once the idea of a project is conceived, proponents idealize it, and plan its phases. This is a crucial moment, since important decisions are taken, which may impact the future of the project, including significant modification in location and technology. The following scheme demonstrates the life cycle of projects (figure 4.1):

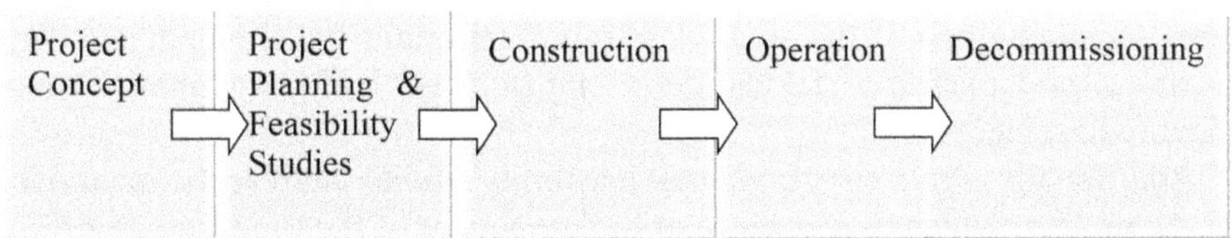

Figure 4.1 Project's Life Cycle. *Source*: International Finance Corporation, *Stakeholders Engagement: A Good Practice Handbook for Companies Doing Business in Emerging Markets* (Washington, DC: International Finance Corporation, 2007), http://www.ifc.org/wps/wcm/connect/641d3c0048855b7c890cdb6a6515bb18/PartTwo_FeasibilityStudies.pdf?MOD=AJPERES.

The planning phase of certain projects, especially large-scale ones, includes feasibility studies. Normally these studies evaluate technological and implementation costs. They also carry out project risk analysis, which assists project developers evaluate the cost-benefits of project implementation and decide whether they will pursue it or not.[2] Considering that environmental matters can be a reason to abandon the implementation of a project, it is a trend (at least for large-scale projects) in Brazil to carry out environmental feasibility studies.

Ideally, social and environmental impacts should be considered in conjunction with other aspects such as economic and technological. The consideration of social and environmental impacts in the planning phase of projects is a practice oriented by sustainability values, which might assist planning the costs and time of its implementation.[3] Realization of environmental feasibilities studies plays an important role in this regard.

FEASIBILITY STUDIES

In some countries, the development of a preliminary study is required to determine whether an EIA should be carried out or not, such as in the United States. This is part of the screening criteria. This section refers to environmental feasibility studies that have been voluntarily carried out by Brazilian enterprises before filing an environmental permit request at the environmental agency.

Environmental feasibility studies aim to provide information on economic, social, and environmental possibility of execution of a project.[4] In general terms, it is a brief analysis of the location considering available technology, legal aspects, and financial resources as well as social and environmental characteristics of the area in light of the proposed project. Examples of actions carried out during environmental feasibility studies are site visits, brief social analysis of the area, identifying important stakeholders and their interests, and planning stakeholder's engagement, including financial and legal requirements.[5]

Additionally, some environmental feasibility studies analyze the main risks of the project, which give a sense to the proponent of the time frame of project approval by administrative agencies. Some companies only carry out an economic feasibility study. However, at least major projects in Brazil develop an environmental feasibility study in order to acknowledge beforehand all the main concerns related to the project.

This is a precautionary practice not only from the environmental point of view but also from economic aspects, since it assists project proponents in financially planning their activity and the main shortcomings they might face. Moreover, an effective environmental feasibility study may assist defining the scope of the EIA. It might also assist the identification of important stakeholders and delimitation of the area of influence of the project.

Brazilian legislation requires proponents of hydroelectric dams to carry out technical and economic feasibility studies prior to its participation in the electricity bid.[6] An environmental feasibility study is not required; instead, the EIA is developed.

An interesting initiative was put into practice by the Brazilian National Department of Infrastructure Transportation (in Portuguese *Departamento Nacional de Infraestrutura de Transporte* or DNIT). In 2007, the DNIT issued an executive order, the Portaria DNIT No 1.705, of November 14, 2007,[7] which requires that major infrastructure transportation projects (above R$20 million or US$5 million) will only be bid after the development of economical, technical, and environmental feasibility studies, which precede the EIA.[8]

The IFC's Stakeholder Engagement Guide encourages companies to carry out stakeholders' engagement at a very early stage of a projects' life cycle, including during the feasibility studies. The main concern of project proponents is the fact that they might not yet have all the answers to some questions or concerns. Furthermore, the knowledge of construction of a project in certain areas might cause speculation, which affects the real estate market as well as prices of goods. However, the IFC's guide points out that, whether it is stated upfront that they are still in the initial phase of planning the project, it is unlikely that stakeholders will act negatively.[9]

Once the environmental feasibility studies are prepared, the project proponent might not have all the answers—it is not a full impact assessment, but some basic information is known. This is the appropriate moment to start stakeholders' engagement. In relation to speculation, once the project proponent files a request for an environmental permit, the news itself of the project raises people's concerns and speculation. Therefore, it is wise to explain to communities the plans of implementing a project rather than leaving it to a much later stage.

An environmental feasibility study normally describes the project, the environmental baseline, the environmental permitting procedure, the time frame it would normally take, the necessity of carrying out an EIA, and the main legislation related to the project. Considering that environmental feasibility studies is a tool that benefits proponents and it is early enough in the life cycle of projects to start engaging stakeholders, specific projects, such as large-scale projects or those planned in sensitive areas—including those in the Amazon rainforest or those that would affect indigenous communities' traditions and/or their lands, among others—should be required to carry out environmental feasibility studies.

The common practice of carrying out environmental feasibility studies should be added to the first step of the EIA procedure. In this case, the study would be filed in conjunction with the preconstruction permit requirement. The environmental permitting agency would then attach the environmental feasibility studies to the EIA to allow public consultations and comparison of both studies. Additionally, to guarantee the quality of feasibility studies, each study should be regulated by the Brazilian government as a short study containing social, environmental, and technological aspects of proposed projects.[10]

This measure would anticipate key issues to consider in EIAs, assisting the development of its scope and mapping stakeholders to participate in the scoping of the issues to be covered in the EIA.

ENGAGING IN THE EIA PROCEDURE

Screening

Screening is the first step of the EIA procedure. It signifies the evaluation of whether a project, policy, or program should be submitted to an EIA. Well-defined legislation, which clearly guides project proponents and governments on when an EIA is needed, demonstrates a country's commitment to sustainable development.[11] According to Rajaram, "The screening guidelines for EIA reflect the first level of commitment the nation displays towards tightening its environmental protection system."[12]

EIA screening may vary among countries. In general, there are three approaches: (1) A list of projects, which can be inclusive or positive lists or exclusive or negative lists. This approach is known as development-centered or threshold. It takes into consideration type and capacity of projects to trigger the necessity of carrying out an EIA.[13] (2) A case-by-case analysis that is also known as the environmental-centered approach. It evaluates the environmental impacts of projects individually regardless of size/capacity. (3) A hybrid approach is often applied.

Regarding the threshold or list of projects, two types of lists are common: inclusion or positive lists and exclusion or negative lists. Countries like Brazil and Thailand adopt the inclusion list, which means that projects listed will undoubtedly be subject to EIA procedure due to their known environmental impacts. Additionally, lists may also have conditions, such as scale and capacity; for instance, a power plant that will generate above 10 MW of electricity must be subject to EIA procedure.[14] As the name suggests, projects and activities listed on the exclusion list will not be subject to EIA procedure because it is already known that they cause minor impacts on the environment.[15] For instance, in India a mining project in an area smaller than 5 hectares is excluded from EIA.[16]

The United States has adopted the case-by-case analysis. The federal agency responsible for developing the project or issuing the environmental permit evaluates whether the proposed project is likely to cause impacts that will significantly affect the environment.[17] In general, and in brief terms, the federal agency has the discretion to determine that a certain project is unlikely to cause significant environmental impact. In these cases, the agency issues a Categorical Exclusion (CE).[18]

The CE means that such project does not need to be submitted to an EIA due to its negligible or minor environmental impacts. However, if federal agencies do not have enough information or if they are not certain about the possible impacts of a project, they might carry out a previous and general analysis: the Environmental Assessment (EA).[19]

The EA will determine whether the project is likely to have significant environmental impacts. The outcome of such analysis may be the determination of an EIA or a Finding of No Significant Impact (FONSI), which exempts the federal agency from developing an EIA for the specific project.[20]

Both approaches (list-of-projects or case-by-case analysis) provide benefits and shortcomings. A common criticism of the list-of-projects approach is that lists are rigid and apply equally to all projects. Thus, some projects included on positive lists (mandatory EIA), although it is already known that they will cause minor environmental impacts, still have to develop an EIA, which would be a waste of time and financial resources.[21]

Some projects or activities have minor impacts when considered individually; however, they might have significant environmental impacts considering other circumstances. For instance, in Indian legislation, the cumulative impacts of several small-scale textile industries in the same region, which were not submitted to EIA due to their minor impacts, will be significant when considered in conjunction.[22] Conversely, Glasson points to strengths of lists, stating that they are simple and quick to apply. Additionally, they generate consistent decisions in different locations and projects.[23]

The main criticism of case-by-case analysis is that it allows a high level of discretion to the authority responsible for evaluating the necessity of requesting EIAs. Therefore, it may lead to poor decisions or abuse of power. However, Glasson states that case-by-case analysis is flexible and therefore it is easier to evolve the procedure.[24]

Considering these strengths and weaknesses, a hybrid approach is more appropriate—such as the system adopted in the European Union on the EIA Directive.[25] It places a "balance between few and abundance of discretionary power,"[26] especially in developing countries, where there is a need for economic growth to fight poverty.[27] Thus, carrying out EIAs is generally seen as a barrier that impedes or delays development, which results in regulations that are weak and/or allow a great level of discretion to authorizing agencies.

Wood asserts that "in practice screening in developing countries is weak because environmental agencies have little power."[28] In fact, the pressure for development from other sectors of government leaves no choice to environmental agencies than authorizing construction and operation of projects without a deep analysis of the need of EIA. Other factors also contribute to this situation, such as lack of qualified personnel in authorizing agencies to review EIAs. Moreover, corruption is a serious and unresolved problem in many developing countries. A well-defined regulation is necessary in order to cut these edges.

Considering the steps of Brazilian EIA procedure, CONAMA Resolution No. 001/86 lists sixteen projects that must be subjected to EIA. Article 2 reads, "The following projects will depend on the development of Environmental Impact Study and its respective Environmental Impact Statement . . . and the permitting of activities which modify the environment, such as: I- roads with two or more lanes; II- railroads. . . ."[29]

The list in CONAMA Resolution No. 001/86 is an inclusion list as well as an illustrative list. It is a unanimous understanding in the literature that the expression "such as" demonstrates that the projects listed are just examples of projects that must develop an EIA.[30] Therefore, projects that are not listed will be assessed by the environmental agency for whether they need to develop an EIA or not. Since only sixteen projects are listed, a large level of discretion is left in the hands of the Federal Environmental Agency or State Environmental Agencies, which complies with the CONAMA Resolution No. 001/86.

Within Brazilian legislation there is not a clear definition of how the environmental agency will take the decision on requesting or waiving a project, not listed in the referred Resolution, from carrying out an EIA.[31] In this regard, the United States has an explicit procedure. The CEQ regulation sets forth that the agency will carry out an EA in order to evaluate whether an EIA is needed. The only Brazilian legislation that contains a vague provision in this issue is CONAMA Resolution No. 237/97, Article 3, which reads, "if the environmental agency verifies that the activity or entrepreneurship will not potentially cause significant environmental degradation, it will define the necessary environmental studies to the environmental permitting procedure."[32] Neither the resolution nor any other legislation provides the meaning of "significant environmental degradation" nor which procedure/method of evaluation the agency will utilize to take such decision.

Beltrão states that the Brazilian Federal Constitution, Article 37 sets forth that government officials will develop their activities guided by principles of legality, impartiality, morality, publicity, and efficiency.[33] Moreover, Law No. 9784 of January 1, 1999, which regulates federal administrative procedures, restates those principles and also adds proportionality and reasoning principles.[34] Therefore, according to Beltrão, arbitrary administrative decisions, for which the agency did not provide the reasons , are subject to judicial review.[35]

Thus, the evaluation of the environmental agency on a certain project regarding the significant environmental degradation should be taken based on scientific and environmental studies and other available data. Such decisions will be part of an administrative procedure containing reasons, and is accessible to the general public, according to Law No. 10650/03, which regulates public access to governmental documents.[36] Otherwise, such decision is considered arbitrary and might be subject to judiciary review.

This interpretation is legitimate and seems to be according to the intent of the Brazilian legislation. However, in practice, environmental agencies have a great amount of administrative procedures, including environmental permitting. Furthermore, decisions are normally not easily available to the public, and in many cases, decisions on the impacts of a certain project are not well-reasoned. Furthermore, Brazilian courts have a large number of lawsuits filed yearly.[37] It is understandable that a solution to a lack of clear explanation on an administrative decision would be filing a lawsuit. However, a clear legislation on screening, establishing the criteria which would define which projects would have significant impacts, would bring enormous benefits to people and the environment.

Therefore, the CONAMA Resolution should provide clear definition of projects which must be subject to EIA under certain conditions, such as capacity, scale, environmental and social impacts, and localization. A hybrid approach similar to the EU EIA Directive 2014/52/EU would narrow the discretionary power of the environmental agency; however, it would still leave some level of discretion, which is necessary in certain circumstances. For instance, in cases where cumulative impacts of several small-scale projects would have destructive social and environmental effects or when projects propose to utilize new technologies.[38]

The EU EIA Directive was enacted in 1985 (85/337/EEC) and has been amended three times (1997, 2003, and 2009). The most recent amendment came into force on May 15, 2014, and was codified in Directive 2014/52/EU. The latest amendment included changes in the screening procedure, making it simpler than before and mandating that decisions must be duly reasoned.[39] As far as screening is concerned, the EIA Directive 2014/52/EU consists of the following: Annex I sets forth mandatory EIA procedure for the activities/projects listed in it while projects listed in Annex II are subject to discretion of the authorizing agency.[40] The agency will carry out a case-by-case analysis taking into consideration criteria stated in Annex III. The amendments to EU EIA Directive included Annex IIA, which lists the information developer is required to provide the authorizing agency.

Some of the examples of Annex I projects include

1. Crude-oil refineries (excluding undertakings manufacturing only lubricants from crude oil) and installations for the gasification and

liquefaction of 500 tons or more of coal or bituminous shale per day. 2. (a) Thermal power stations and other combustion installations with a heat output of 300 megawatts or more; (b) Nuclear power stations and other nuclear reactors including the dismantling or decommissioning of such power stations or reactors (1) (except research installations for the production and conversion of fissionable and fertile materials, whose maximum power does not exceed 1 kilowatt continuous thermal load).[41]

The following are some of the examples of Annex II:

1. Agriculture, Silviculture and Aquaculture:
   (a) Projects for the restructuring of rural land holdings;
   (b) Projects for the use of uncultivated land or semi-natural areas for intensive agricultural purposes;
2. Extractive Industry
   (a) Quarries, open-cast mining . . .
   (b) Underground mining.[42]

Some examples of information required in Annex II.A are
1. A description of the project, including in particular:
   (a) a description of the physical characteristics of the whole project . . .
   (b) a description of the location of the project[43]

Examples of criteria to determine whether the project listed in Annex II should be subject to an EIA as listed in Annex III:

2. Characteristics of projects. The characteristics of projects must be considered, with particular regard to: (a) the size and design of the whole project; (b) cumulation with other existing and/or approved projects; (c) the use of natural resources, in particular land, soil, water and biodiversity; (d) the production of waste; (e) pollution and nuisances; (f) the risk of major accidents and/or disasters which are relevant to the project concerned, including those caused by climate change, in accordance with scientific knowledge; (g) the risks to human health (for example due to water contamination or air pollution).
3. Location of projects. The environmental sensitivity of geographical areas likely to be affected by projects must be considered, with particular regard to: (a) the existing and approved land use; (b) the relative abundance, availability, quality and regenerative capacity of natural resources (including soil, land, water and biodiversity) in the area and its underground; (c) the absorption capacity of the natural environment.[44]

This book proposes a hybrid approach, similar to the EU EIA Directive, to select which projects would go through the EIA containing a robust public participation process. To put it simply, Annex I projects would require the EIA, including the full public participation process, while Annex II projects would be subject to discretion of the environmental agency, which could request an EIA. However, it would have a simpler public participation process. Thus, there would be two sets of EIA procedures: the Annex I procedure, which would require the EIA full public participation, and the Annex II EIA, requiring simpler procedure.[45]

The long or short procedures have more obligatory steps in public participation or least actions in this regard. This is necessary because some projects due to complexity, localization, size, and involvement of different stakeholders require more public engagement than others. Furthermore, in order to have meaningful public participation it is necessary to educate and empower communities. Therefore, some specific measures are necessary to achieve this goal.

Adapting this proposal to the Brazilian reality consider the Volta Grande Project, mentioned in the previous chapter. It will affect several different communities, such as fishermen, artisanal miners, riverbank, among others. This project planned to be implemented in a sensitive area involving different stakeholders should carry out an engagement of stakeholders and develop a set of public participation process from the very beginning of the EIA procedure. In other words, a full Public Participation EIA procedure would be necessary in this case.

This book recommends the longest procedure be implemented in Brazil. The Annex I list should include localization, size, and complexity conditions, such as hydroelectric dams which would produce above certain amount of electricity; projects in the Amazon region or in other sensitive biomes, such as the Pantanal; and medium projects, which could have cumulative impacts that would seriously compromise the social and environmental quality of surrounding communities.

The hybrid screening criteria would be appropriate to select the projects that should develop an EIA and would require a more robust public participation process, including the requirement of environmental feasibility studies.

CONCLUSION

Social and environmental considerations should be integrated into the life cycle of projects. Such a measure would forecast the main challenges related to the approval of projects such as time frame of environmental permitting and EIA procedures due to the complexity of the project and social and environmental concerns related to the specific site planned to implement the project.

The environmental feasibility studies that have been carried out as a voluntary practice by major projects in Brazil is a brief assessment which assists project proponents in forecasting the main socio-environmental matters as well as provides a basis for the identification of stakeholders within the area of influence of the project.

Furthermore, the environmental feasibility studies are a starting point to hold communities' discussions on the scoping of issues to be included the EIA. Therefore, such practice should be incorporated as an obligatory first step in the EIA procedure for specific projects, according to the screening criteria established in the regulation.

When project proponents engage in the Brazilian EIA, the first step of this procedure is the screening. The inclusive list of projects in CONAMA Resolution No. 001/86 is illustrative and therefore great discretionary powers are left to environmental agencies to decide whether a project requires an EIA or not.

Scholars state that public officials must act accordingly with constitutional principles, including impartiality and proportionality. Otherwise, decisions can be revised in court. However, legislation clarifying which projects should develop an EIA according to their characteristics would assist practitioners and diminish the need of searching legal remedies.

Furthermore, the hybrid approach, similar to the EU Directive provides a balance between positivism and discretionary powers. This approach would be the screening of projects which are most sensitive and therefore require more public participation in all their phases as it will be seen throughout this book.

NOTES

1. John Glasson et al., *Introduction to Environmental Impact Assessment*, 3rd ed. (London: Routledge, 2012), 15.
2. International Finance Corporation, *Stakeholders Engagement: A Good Practice Handbook for Companies Doing Business in Emerging Markets* (Washington, DC: International Finance Corporation, 2007), http://www.ifc.org/wps/wcm/connect/641d3c0048855b7c890cdb6a6515bb18/PartTwo_FeasibilityStudies.pdf?MOD=AJPERES.
3. Glasson et al., *Introduction to Environmental Impact Assessment*, 15.
4. A. Urkiaga et al., "Methodologies for Feasibility Studies Related to Wastewater Reclamation and Reuse Projects Desalination," *Desalination* 187 (February 2006): 263–269.
5. International Finance Corporation, *Stakeholder Engagement.*
6. "Belo Monte," Aneel, last accessed October 22, 2019, http://www.aneel.gov.br/aplicacoes/hotsite_beloMonte/index.cfm?p=7.

7. Portraria DNIT No. 1.705, de 14 de Novembro de 2007, Diário Oficial da União (D.O.U.) de 20.11.2007 (Braz.).

8. The requirement for Economic, Technical, and Environmental Feasibility studies refers to infrastructure projects of waterways, roadways, and railroads. *See* Departamento de Infra-estrutura de Transporte (DNIT), *Estudos de Viabilidade Técnica, Ecônomica e Ambiental* (Technical, Economic and Environmental Feasibility Study, Department of Infrastructure and Transportation—DNIT), last accessed March 3, 2016, http://www.dnit.gov.br/download/planejamento-e-pesquisa/planejamento/estudos-de-viabilidade/lcs-097-2010-solic-public-texto-evtea-site-dnit.pdf.

9. International Finance Corporation, *Stakeholder Engagement.*

10. Li-Yin Shen et al. state that "governments have an important role in promoting sustainability of construction project at the stage of project feasibility study. The government should guide with policies, laws and regulations. This role should be practiced through various ways including laws and regulations, industrial specifications, administrative examination and approval, tax fine and other means." Li-Yin Shen et al., "Project Feasibility Study: The Key to Successful Implementation of Sustainable and Socially Responsible Construction Management Practice," *Journal of Cleaner Production* 18, no. 3 (February 2010): 254–259.

11. T. Rajaram and Ashutosh Das, "Screening for EIA in India: Enhancing Effectiveness Through Ecological Carrying Capacity Approach," *Journal of Environmental Management* 92, no. 1 (January 2011): 140–148.

12. Ibid.

13. Ibid.

14. Resolução CONAMA No. 001, de 23 de Janeiro de 1986, art. 2, XI, Diário Oficial da União (D.O.U.) de 17.02.1986 (Braz.).

15. Glasson et al., *Introduction to Environmental Impact Assessment*, 90.

16. Rajaram and Das, "Screening for EIA in India," 140, 142.

17. "Environmental Impact Assessment," Environmental Protection Agency, last accessed October 22, 2019, http://www.epa.ie/monitoringassessment/assessment/eia/.

18. Council on Environmental Quality (CEQ) Regulations for Implementing the Procedural Provisions of the National Environmental Policy Act, 40 C.F.R. § 1508.4 (2015).

19. CEQ regulations provide that the public should be involved as much as possible in the EA. (40 C.F.R. § 1508.9). The agency shall involve environmental agencies, applicants, and the public, to the extent practicable, in preparing assessments required by 40 C.F.R. § 1508.9(a) (1).

20. 40 C.F.R. §1501.4(e) (2015).

21. Paulo Cesar Gonçalves Egler, "Improving the Environmental Impact Assessment Process in Brazil" (PhD Dissertation, School of Environmental Sciences University of East Anglia, 1998), 67, http://busca.ibict.br/SearchBDTD/search.do?command=search&q=+instituicao_defesa:%22University%20of%20East%20Anglia%22.

22. Rajaram and Das, "Screening for EIA in India," 140–146.

23. Glasson et al., *Introduction to Environmental Impact Assessment*, 91.

24. Ibid.

25. Directive 2014/52/EU of the European Parliament and of the Council of 16 April 2014 Amending Directive 2011/92/EU on the Assessment of the Effects of Certain Public and Private Projects on the Environment, 2014 O.J. (L 124) 1.

26. Christopher Wood, "Environmental Impact Assessment in Developing Countries: An Overview" (paper presented at New Directions in Impact Assessment and Development: Methods and Practice, November 24, 2003), http://www.academia.edu/3420793/Environmental_impact_assessment_in_developing_countries_an_overview.

27. Rajaram and Das, "Screening for EIA in India," 141.

28. Wood, "Environmental Impact Assessment in Developing Countries."

29. Resolução CONAMA No. 001/86, art. 2.

30. Paulo Affonso Leme Machado, *Direito Ambiental Brasileiro* (Brazilian Environmental Law), 21st ed. (São Paulo: Malheiroso Editores, 2013).

31. Antonio F.G. Beltrão, *Aspectos Jurídicos do Estudo de Impacto Ambiental (EIA)* (Legal Aspects of Environmental Impact Assessment [EIA]) (São Paulo: MP Editora, 2008).

32. Ibid.; Resolução CONAMA No. 237, de 19 de dezembro de 1997, Diário Oficial da União (D.O.U.), art. 3, de 22.12.1997 (Braz.).

33. Beltrão, *Aspectos Jurídicos*; Constituição Federal (C.F.) (Constitution) art. 37 (Braz.).

34. Lei No. 9.784, de 29 de Janeiro de 1989, Col. Leis Rep. Fed. Brasil, 1, 41, Fevereiro 1999 (Braz.). The reasoning principle means that all administrative decisions must be justified.

35. Beltrão, *Aspectos Jurídicos*; Constituição Federal (C.F.) (Constitution) art. 37 (Braz.).

36. Beltrão, *Aspectos Jurídicos*; Constituição Federal (C.F.) (Constitution) art. 37 (Braz.); Lei No. 10.650, de 16 de Abril de 2003, Diário Oficial da União (D.O.U.) de 17.04.2003 (Braz.).

37. In 2013, the number of legal actions filed at state courts reached 20,282,181. Tribunal de Justiça do Estado de São Paulo, *Processos Distribuídos Justiça Estadual Brasileira—Dados de 2009 a 2013* (Lawsuits Filed at the State Courts—Data from 2009 to 2013) (January 2015) (on file with author).

38. Directive 2014/52/EU, 2014 O.J. (L 124) 1.

39. Ibid.

40. Annex II is included in Directive 2011/92/EU, 2011 O.J. (L 26) 1.

41. Directive 2011/92/EU, 2011, Annex I, O.J. (L 26) 1.

42. Ibid., Annex II.

43. Directive 2014/52/EU, Annex IIA.

44. Ibid., Annex III.

45. The Netherlands adopted a similar system; however, it may differ in its application. There are the "Environmental Impact Assessment for complex decisions" and the simplified procedure for relatively simple permitting procedures. See "Dutch Legislation," Netherlands Commission for Environmental Assessment, last accessed January 14, 2017, http://www.commissiemer.nl/english/our-services/types-of-advisory-report.

Chapter 5

Scoping

Scoping is the next step in the EIA procedure. Once established that an EIA is required for certain project (through a list or a case-by-case analysis), it is determined, in general by the EIA reviewing agency, what topics the EIA will cover. Scoping is therefore subjective to the project and its location. For instance, assessment of project impacts on mangrove, traditional communities, and underground water is only possible in areas with these characteristics. In the same vein identification of communities' views and concerns differ depending on the project and the location.

The scoping means deciding on which significant impacts should be assessed in light of the proposed project. The question is, what is "significant" in the view of affected communities and experts? As Weston states, scoping is a "heart human centered approach."[1] It relies heavily on the understanding of "significance." Thus, public participation and consultation are so important in this phase.

Generally, scoping outlines environmental baseline information, alternatives to the proposed project, and a selection of impacts to be evaluated.[2] The UK Department of the Environment states that "scoping is the key to a good quality environmental impact statement."[3] Best practice indicates a set of principles to be considered in the scoping of an EIA, which are the following:[4]

- Early stage
- Engagement of key stakeholders and open process
- Consider alternatives of project design and location
- Focus on key social and environmental issues
- Definition of clear limits (spatial, temporal, socioeconomic)
- Propose monitoring of impacts throughout the life of the project

Scoping shapes the EIA focusing on the most important impacts and leaves aside the least relevant ones.[5] This is important because the EIA is a complex process which involves collection and analysis of data, and participation of different stakeholders, which might take a long period of time.[6] Therefore, a focused and planned EIA has more chances to achieve its purpose, which is to reconcile social, environmental, and economic aspects in economic decisions, than other EIAs which do not have the same features.

Some legislation determines the scoping as an obligatory phase of the EIA process, such as in Canada, United States, and South Africa.[7] Conversely, others do not contain such obligation. For instance, Article 5 of the EU EIA Directive lists the minimum content of an EIS and Paragraph 2 sets forth that

> Where requested by the developer, the competent authority shall issue an opinion on the scope and the level of detail of the information to be included by the developer in the environmental impact assessment report. The competent authority shall consult the authorities referred in Article 6 (1) before it gives its opinion. Member States may also require the competent authorities to give an opinion as referred to in the first subparagraph, irrespective of whether the developer so requests.[8]

In other words, the proponent of the project will elaborate the scope. The competent authority will only issue an opinion on the scope if requested to do so.

Although there are thresholds that guide the development of an EIA scope, such as lists (Western Australia) and the OECD guidelines, there is great level of political decision in this phase.[9] In this regard, Weston carried out a survey on methodologies used to decide the scope of EIAs. Thirty-three consultancies replied; among those 50 percent said that experience and judgment were key factors to consider on which impacts would be covered in the EIA.[10] According to Weston this is not a drawback. It is a strength of scoping since the significance of impacts is a relative concept, which depends on different projects, environmental impacts, and affected people.[11]

ENVIRONMENTAL AND SOCIAL SCOPING: AN INTEGRATED APPROACH

EIA scoping includes (or should include) both ecological concerns and social impacts of projects.[12] In this regard, Bealands and Duinker propose two scoping approaches: social and ecological, where the first focuses on identifying and understanding the values of different groups and how such values can be reflected in the EIA.[13] The ecological approach sets the guideline on the natural resources and other issues to be evaluated in the EIA. Although such separation is didactic, the assessment of impacts should consider an integrated approach in which social and ecological impacts are evaluated in conjunction. For instance, the relationship of indigenous communities with the land on which they live and the impacts on the soil cannot be assessed independently because such impacts are interrelated.

Having said that, scoping should encompass not only the scientific aspects of projects' impacts but also the more diffuse considerations to be addressed, such as the impact on the quality of life in an area, levels of crime, and community health.[14] Affected communities' concerns should be addressed on the scope of the EIA, such as fear of losing their houses in case of relocation, concerns of losing their ties within the community and with the space they occupy. In this regard, there are some practices on social impact as a separate evaluation of the EIA,[15] which, although it is well-recognized as best practice, it is still not very disseminated.[16]

For the purpose of this book, social and environmental impacts of projects should be analyzed under the auspices of the EIA procedure because these two dimensions cannot be assessed independently, and also because the book proposes measures to improve public participation in Brazil, a country which presents many deficiencies in EIA. Thus, proposing a mandatory separate Social Impact Assessment, at least at present, would not work in practice.

Considering that scoping aims to select the significant impacts of projects, it is crucial to consider public input in order to identify what is "significant in the view of the people who will have to live with the project and not just the 'experts' who will not."[17] However, involvement of key stakeholders and public participation in the scoping is a major weakness in many countries.[18] For instance in the UK, research denotes that there is limited community consultation, due to lack of legal requirement.[19] Additionally, a research carried out on twenty scoping reports from Portugal demonstrated that effective public participation was not carried out in this phase of environmental assessments.[20]

Scholars and practitioners have recognized the main obstacles of having an open and participatory scoping.[21] Some enterprises see it as one more step which will make the EIA procedure longer and expensive. For instance, a research carried out by the World Resource Institute, during the period of 2002–2005, noted that among twenty-nine projects analyzed worldwide, only one-third of the cases had some form of public participation on scoping, yet they were not considered "strong" participation.[22]

Furthermore, it is difficult to reconcile all different views in particular on implementation of large projects.[23] However, the strength of having an open scoping benefits especially controversial entrepreneurship because it builds public confidence concerning the project proponent, the government, and the competent authority responsible for conducting the EIA process.[24]

An open scoping might reduce costs because it will address problems identified by the public at an early stage of the EIA process. Therefore, it will diminish attempts to halt the project, clarifying public doubts and common concerns by building a long-standing relationship with communities.[25] There are several examples of delays in the authorization of projects due to lack of public participation at early stages of EIAs.

Sanchez cites the proposition of a hydroelectric dam in São Paulo, Brazil. The main problem of this case was that the alternative selected by the project proponent would deviate the waters of the Paranapanema River reducing its flow at the urban portion.[26] The community that lived in the area of the proposed project strongly objected to it because the river was seen as essential to their lives and it was part of the landscape of the city. As a result, the project proponent had to present three different versions of the EIS, which delayed the authorization of the project and consequently made the project more expensive. The community expressed their dissatisfaction with the possibility of a reduction in the water flow of the river and achieved great modifications to the project. Sanchez states that early public participation would have denoted the public concern earlier and a better project alternative would have been selected avoiding costs and delay.[27]

Public participation in EIA scoping provides a basis for social control. It serves as a guideline for the public to compare the draft EIS with what was proposed on its scope. Affected communities will have truer input and meaningful participation if they do so by having their views addressed at the planning stage of the EIA procedure. An early participation has more chances to affect the results of decision-making on developmental projects.[28] Not only affected communities but also governmental agencies will be surprised with an EIS in which they did not participate and consequently did not cover all the relevant issues regarding the proposed project.[29] The Volta Grande Project is an example of this situation. Thus, the scoping should be seen as a "participative and collaborative phase" of EIA, which might take longer; however, it will avoid later complaints and project paralysis.[30]

Some norms request public participation in scoping. Others leave to the competent authority's discretion to open the EIA scope for public participation or not. For instance, the United States CEQ regulations 1997 recognize the necessity of having an open and early process during the EIA scoping as well as the involvement of other actors in the process. Section 1501.7 of the referred regulation sets forth the following: "a) As part of the scoping process the lead agency shall: Invite the participation of affected federal, state, and local agencies, any affected Indian tribe, the proponent of the action, and other interested persons (including those who might not be in accord with the action on environmental grounds)."[31]

Countries such as Belgium (Brussels and Walloon region), Denmark, Finland, Netherlands, Spain, and Sweden have open EIA scoping processes in which the public consultation is obligatory.[32] In Canada, the draft Tailored Impact Statement Guidelines, which contains the scope of the issues to be covered in the EIA, is posted on the registry for public comments. After comments are received, the Impact Assessment Agency finalizes the guidelines and provides it to the project proponent.[33]

Conversely, in Austria, Germany, Ireland, and the UK, the competent authority decides whether or not the EIA scoping will be open to public participation.[34] However, relevant authorities or agencies are always consulted.[35] More specifically, in the UK, (England and Wales), if the EIA developer requests to the competent authority a scoping opinion, it will be available for the public on the planning register for a period of two years.[36]

Overall, scoping is obligatory and open to public participation in some jurisdictions or it is under the authorization agency's discretion to decide whether it will open the scoping of an EIA to public input. In some countries scoping happens just between developers and authorization agencies, as this is the case in Brazil.

In Brazil, CONAMA Resolution No. 001/86 does not state that scoping is an obligatory phase of the EIA, as it is further explained in the next sections.

TERMS OF REFERENCE (TOR)

Ideally, scoping results in TOR, which might be approved by the authority responsible for conducting the EIA process.[37] In Brazil, Articles 5 and 6 of CONAMA Resolution No. 001/86 delimitate the minimum content of an EIS. However, it does not specify who will be responsible for preparing the TOR. Furthermore, it does not mention the necessity of involving the public and other interested agencies in the preparation of this document.

> Article 5: The environmental impact study . . . will follow the general guidelines: I—Contemplate all technological and locational alternatives to the installation of the project, comparing them with the non-realization of the project; II—Identify and assess systematically the environmental impacts during the installation and operation of the project; III—Define the geographical limits to be directly or indirectly affected by the impacts, which will be the area of influence of the project . . .; IV—Consider the governmental plans and programs proposed or in implementation in the area of influence of the project and its compatibility.[38] Article 6: The Environmental Impact Assessment will develop,

> at least, the following technical activities: I—Environmental Diagnosis of the area of influence of the project. Complete description and analysis baseline information regarding the environmental resources of the area and its interactions to characterize the environmental situation of the area, before the construction of the project . . . II—. . . III—Definition of the negative impacts mitigation measures . . . IV—Development of the monitoring program. . . . Paragraph—In determining the realization of the Environmental Impact Study, the competent environmental agency, or IBAMA, or when it is possible, the municipality will provide additional instructions, when necessary, according to the projects' peculiarities and the environmental characteristics of the area.[39]

In 2011, the Environmental Ministry edited several ministerial ordinances,[40] which regulate EIAs for certain projects. Interministerial Ordinance No. 60/15 (which revoked Interministerial Ordinance No. 419/11) regulates the participation of some governmental agencies in the environmental permitting procedure;[41] Interministerial Ordinance No. 421/11 regulates the environmental permitting procedure and environmental regularization of transmission lines, at the federal level;[42] Interministerial Ordinance No. 422/11 regulates the environmental permitting procedure of offshore and transition zone land-sea oil and gas exploration and exploitation activities.[43]

Interministerial Ordinance No. 60/15 (Article 2) states that TOR is a document that should be elaborated by IBAMA and establishes the necessary content of environmental assessments.[44] It also regulates the issuance of specific TOR by certain governmental agencies. An analysis of Interministerial Ordinance No. 60/15 is presented further in this section.

Interministerial Ordinance No. 421/11 (transmission lines) states that IBAMA will issue the TOR. The project proponent might participate in its development, if he/she requests so.[45] It also establishes deadlines for TOR issuance, verification of compliance between the EIS and the TOR, and limit of time in which TOR is valid. The referred Ordinance No. 421/11 provides TOR models for EIA and other environmental studies in its annexes.[46]

In the same vein, Interministerial Ordinance No. 422/11 (oil and gas activities) establishes deadlines and other requirements concerning the elaboration of the TOR.[47] It is important to note that it determines that IBAMA must post the TOR on the internet, in a portal created specifically for this aim.[48] As it was demonstrated, the TOR is reasonably well-regulated in the EIA process concerning transmission lines and oil and gas exploitation and exploration. Conversely, all other activities, which must comply with CONAMA Resolution No. 001/86, do not have the same range of details. It is interesting to note that for offshore activities, there is the possibility of accessing the TOR online, which represents an attitude of transparency and allows social control. However, in terms of social impacts offshore oil and gas exploration and exploitation have fewer impacts in comparison with inland activities (mining for instance). Therefore, a general regulation on the elaboration of TOR, including the opportunity for public participation, is still missing in the Brazilian legal system.

At the state level, few states have regulations on the preparation of TOR.[49] In Bahia, CEPRAM Resolution No. 2929/02 establishes the realization of a public hearing before the elaboration of the TOR,[50] namely the previous public hearing and another hearing when the EIS is completed. However, experience demonstrates that both public hearings are spaces where the population does not discuss the potential environmental impacts of the proposed project, but mostly talk about their poor life conditions.[51] In Rio de Janeiro, State Law No. 1356/88 has one provision simply stating that the environmental agency will orient the development of the EIA through Technical Instructions.[52]

Although some states have provisions in developing TORs, in practice, the EIA developer, who generally are consultancies, prepares the TOR and consults the environmental agency on its content. The environmental agency makes considerations or accepts it entirely. In Brazil, generally, there is no public input in the TOR. The public participates in the EIA procedure only when the EIS is already completed. And normally the public cannot check whether the EIS is adequate with what was proposed at the very beginning of the process.

Thus, participation happens at a later stage of the process.[53] Consequently, it does not reflect the views of society in the EIS. Additionally, it does not include local or traditional knowledge in the area selected to implement the project. Such knowledge would help the EIA developer to understand the areas and potential impacts. For example, local communities might assist identification of endangered or threatened species of animal or plants, and point out archaeological sites of great historical importance.

Whether local communities provide their own point of view of the geographic areas where they live as well as their standard of living and concerns, this would assist developers in addressing impacts and propose mitigation measures.[54] Community organization for Little Bouctouche River, in Canada, offers a good example; community members organized themselves and offered scientific input as well as their experience as long-standing residents of the margins of the river that cumulative impacts needed to be assessed within the EIA of dismounting a bridge that was probably impacting the river.[55]

Furthermore, there is no real discussion of alternatives to the proposed project because the EIS was already developed and it certainly indicates that the best alternative is the one which is being proposed.[56] Therefore, the fact that CONAMA Resolution No. 001/86 does not set a formal scoping procedure and does not allow for public participation in this phase of EIA weakens the entire process.[57] Beltrão states that public participation at late stages of the project, as it is in Brazil, generates lack of confidence from both sides: the project proponent and the public. The project proponent spends time and financial resources in developing the EIS.[58] However, after public participation, the project design or technology might change considerably. The public feels that their participation is pointless and will not have any true input in the decision-making of development projects because the EIS is already completed.[59]

The EC Guidance on Scoping states that "early consultation with interested parties can be very valuable in avoiding later delays if new issues emerge from consultation only after the EIS is submitted."[60] The same guidance also clarifies that although consultations at the scoping phase are very important, the scope should be flexible enough during the process to amend new issues and information.[61]

Participation of Governmental Agencies in the Development of Brazilian EIS/TORs

The participation of governmental agencies in the EIA is another weakness of this procedure in Brazilian practice. In 2012, the Court of Audit issued a report pointing to a lack of involvement of governmental stakeholders in Brazilian environmental permitting procedure.[62] The report cites the case of the environmental permitting procedure of the northern Brazil hydroelectric power plants, Jirau and Santo Antônio, in the State of Rondonia, which did not involve the participation of INCRA.

Participation of INCRA would be of great importance because in the region where the power plant was slated there are lands which are part of the government's agrarian reform areas. However, there are communities living on these lands. They do not hold formal title but still have been occupying the area for many years. Thus, while there are massive amounts of lands being flooded by dams there are also many people living in un-regularized lands. This is a common problem in the northern region of Brazil. The report also mentions the necessity of participation of other governmental agencies in assisting the assessment of social impacts as a result of the hydroelectric power plants.

Interministerial Ordinance No. 60/15 regulates the participation of FUNAI, the Palmares Cultural Foundation (in Portuguese *Fundação Cultural Palmares* or FCP),[63] the National Cultural and Artistic Heritage (IPHAN), and the Health Ministry. The ordinance determines that if the proposed project impacts areas of interest of the aforementioned agencies, they must participate in the EIA procedure.[64]

On the one hand, Interministerial Ordinance No. 60/15 represents a positive step toward the involvement of stakeholders (governmental agencies) in the EIA procedure, considering that until 2011 there were not any clear guidelines capable of orienting EIA developers and environmental agencies. Another positive aspect is the fact that it provides minimum content of specific TORs to be issued by the referred agencies.[65] For instance, Annex IIA provides examples of TOR of projects in areas endemic to malaria and Annex IIB provides the content of TOR to be observed by FUNAI in the elaboration of a specific TOR which will guide studies on projects' impacts on indigenous communities.[66] On the other hand, the Interministerial Ordinance has been criticized because it presents some arguable parameters to determine the interference of projects on indigenous lands.[67] Article 3 of Interministerial Ordinance No. 60 states that

> IBAMA, at the beginning of the environmental permitting procedure, in the Activity Characterization Form, must request the project proponent information regarding potential interference in indigenous lands . . . and endemic areas or regions to malaria: § 20—. . . the mentioned interference is presumed when: I. On indigenous lands, when the proposed project is planned to be located on their lands or presents elements that may result in direct socio-environmental impact in the interior of the indigenous land, according to the geographical limit established in Annex I.[68]

Annex I establishes geographical limits, differentiating projects in the Amazon (known as legal Amazon), in which the spatial limit is larger, from projects to be located in other regions of the country.[69] In the Amazon, mining projects will presumably interfere in indigenous lands if they are located within 10 kilometers of their lands. However, in other parts of Brazil the limit is 8 kilometers.[70] Therefore, if a mining project is located in the Amazon, up to 10 kilometers away from an indigenous land, the project proponent is responsible for contacting FUNAI, which will probably issue a specific TOR.

As discussed in chapter 3, the establishment of maximum distances provides a parameter for project proponents and EIA developers on the involvement of government agencies; however, there are shortcomings in this approach. Generally, indirect impacts are seen as less important and, therefore, generate less mitigation measures.[71] Moreover, indirect impacts do not require indigenous peoples' free, prior, informed consent, according to the spatial limitation, since "the project is not causing direct impact within the land."[72] However, impacts on indigenous communities are not restricted to impacts that affect their land directly.[73] For example indigenous communities might use the water of a river that also serves the project. Thus, if the quality of water changes because of the projects' activity, it will affect them.[74] Although Ordinance No. 60/15 regulates the participation of some governmental agencies, there are many situations in which agencies are not consulted on the TOR, as demonstrated in the analysis of the Volta Grande Project EIA procedure.

EIA practitioners and affected communities argue that bias is a problem in drafting TORs and EIS from EIA developers. Therefore, the next sections discuss important components of scoping and matters that affect the inclusion of social impacts, public participation, as well as the definition of stakeholders to participate in the EIA at early stages. Accordingly, how should the TOR be elaborated? Who should participate in the TOR and how? How should the main stakeholders to participate in the TOR be identified?

How Should the TOR Be Elaborated?

In some countries, the public is invited to comment on the TOR previously drafted by the EIA developer. Additionally, workshops and public hearings are held to discuss this document.[75] For instance, the EIA procedure of the Ekati Diamond Mine in Canada encompassed scoping meetings in eight communities of the North West Territories and received fifty written comments and heard 125 presenters in order to issue the final guidelines for the EIS.[76] Furthermore, some EIA systems have an independent authority, which coordinates the development of EIA and the EIS revision; Finland and the Netherlands have independent reviewing agencies, as discussed in chapter 8.[77]

Brazil lacks a comprehensive regulation on EIA procedure, including the determination of an open scoping. The public should be consulted during elaboration of the TOR.[78] As was demonstrated in the Ekati Diamond Mine, scoping workshops with affected communities, environmental groups, and other stakeholders are possible and should be implemented.[79] The realization of workshops is of great value because it allows two-way interaction and exchange experiences among participants.

Westman states that, considering different methods of public participation, workshops increase education and involvement of citizens and have great chances of influencing decision-making.[80] In comparison to public hearings, structured workshops are more effective. However, public hearings are also important instruments to disseminate information.

Therefore, a comprehensive framework establishing scoping workshops would engage important stakeholders at the beginning of EIA. Brazil would benefit by adopting a screening criterion considering types of projects and localization in order to determine whether the EIA would require scoping workshops or just written public comments. This is because workshops demand time and financial resources.[81] For instance, projects to be located in the Amazon rainforest should definitely hold scoping workshops during the EIA process because of the peculiarities of the region, the different groups who live there and who are affected by projects and the importance of that ecosystem not only for Brazil but also for the world. In this way, the population would participate in the planning phase of the EIA and could have their views addressed in the EIS.

According to Glasson, "a careful balance needs to be struck between consultation that is early enough to influence decisions and consultation that is so early that there is no real information on which to base any discussions."[82] Thus, in order to support the realization of scoping workshops the presenter should utilize the environmental feasibility studies previously filed at the agency, as discussed in chapter 4.

For more complex projects or projects located in specific locations, the realization of scoping workshops would assist development of the TOR. Once the TOR is finalized, it should be made publicly available for public comments and possibly, depending on each case, the realization of public hearings to present its final version.

In Brazil, one of the main criticisms to public hearings and meetings prior to hearings carried out by project proponents is the fact that these meetings and hearings are conducted by project proponents who have an interest in developing the project; therefore, the way they conduct these events is embodied by partiality and the community sees it as cooptation, especially regarding the previous meetings. Therefore, the realization of workshops and hearings by a neutral facilitator is necessary. Furthermore, in order to guarantee impartiality in the development of TOR and the EIS an independent committee should revise them and it would be a channel of communication with society.

Overall, an important step for change in the Brazilian EIA procedure would be the legal obligation of public consultation during construction of the TOR. Beltrão says that "It is more efficient invite stakeholders to participate at the beginning of the EIA because their comments are essential to define the scope of the proposed project. Public will have more interest in participate considering that they can provide input at early stages of the procedure, rather than a procedure that is nearly concluded aiming stakeholders' confirmation."[83] Thus, this document would address communities' concerns and they could also contribute with their local knowledge. Such requirement would lead to more democratic, transparent, and accountable decision-making.[84]

These measures would diminish the imbalance that currently exists in Brazilian EIA procedures where the power to manipulate the EIA findings is with the project proponent and consultants. Furthermore, the decision-making is entirely in the hands of the environmental agency.[85] Public consultation during the elaboration of TOR would also assist in the identification of stakeholders. For instance, interested governmental agencies which must be invited to participate in the EIA procedure. In the case of the Volta Grande Project, agencies such as Health and Land Reform Area did not participate in the EIA procedure. This situation could have been avoided utilizing techniques such as mapping stakeholders and open scoping workshops.

Therefore, to open the scoping for public participation, it is necessary to target who should participate in the drafting of TOR so important authorities and agencies would not be missed as it happened in the case study. Best practices would involve mapping stakeholders at the scoping phase of EIA to identify the main actors who should be involved in consultations of the TOR and subsequent stages of the procedure.

Who Should Participate in the TOR? How Should the Main Stakeholders Participating in the TOR Be Identified?

According to Glasson, Therivel, and Chadwick, "scoping should begin with the identification of individuals, communities, local authorities and statutory consultees likely to be affected by the project; good practice would be to bring them together in a working group and/or meetings with the developer."[86]

The EC Guidance on Scoping contains a consultation checklist of potentially interested stakeholders.[87] It divides them into three main groups: the environmental authorities, other interested organizations, and the public. The checklist provides examples of each of these categories, such as environmental authorities (i.e., health and safety authorities), other interested authorities (i.e., local employers and business association), and the public (i.e., landowners and residents).[88] Guidelines are useful tools and might assist in the identification and selection of important agencies, organizations, and other interested stakeholders in the project. However, lists are not flexible, and they might miss important actors.

Therefore, threshold criteria of identification of stakeholders would be more appropriate considering different projects and their location. For instance, proposals of massive projects such as large dams in sensitive areas such as the Amazon Forest or other regions, where different social groups might be affected by projects, should map stakeholders to participate in the EIA scoping. This is important because normally, different groups and agencies do not participate at early stages of EIAs.

Mapping Stakeholders

Development projects involve many different stakeholders. Glasson divides stakeholders of major projects into four groups: (1) developers, (2) those directly or indirectly affected by the project, (3) the government and regulatory agencies, and (4) various intermediaries (i.e., consultants, advisers, etc.). The permitting agency has an important role in regulating and managing the relationship among these groups.[89] The identification of all interested parties in the proposed development can be a challenge.[90]

The United States Environmental Protection Agency (US EPA) published a guide containing techniques to assist in the identification of stakeholders that should be involved in development projects. It provides examples of interests that may be important to the community and should be considered by project proponent, such as health, safety, pollution, property values, jobs, congestion, crime, and local economy.[91] In this same vein, the IFC Stakeholder Engagement Handbook provides examples of questions to identify main stakeholders in relation to projects: (1) "Who are the most vulnerable among the potentially impacted, and are special engagement efforts necessary?"[92] (2) "At which stage of project development will stakeholders be most affected (e.g., procurement, construction, operations, decommissioning)?"[93] (3) "Who strongly supports or opposes the changes that the project will bring and why?"[94] (4) "Who is it critical to engage first, and why?"[95] (5) "What is the optimal sequence of engagement?"[96]

Interaction of different agencies and actors might be a complicated task. Glasson defines the EIA, within the process of authorizing a project, as the "conflicting resolution phase."[97] Conversely, EIA is commonly seen as a requirement to be fulfilled in order to comply with the law and obtain project approval. However, the EIA, if well-developed, provides opportunity for discussions and negotiations on conflicting views. The US CEQ Regulation recognizes the necessity of cooperation among agencies and states that cooperating agencies shall participate in the EIA at the earliest time possible. In addition, it allocates responsibilities in analyzing the EIS according to their expertise.[98]

In Brazil, Article 11, § 1 of CONAMA Resolution No. 001/86 states that public agencies which express interest or have direct relation with the project will receive a copy of the nontechnical summary of the EIS for information and provide their input. This is the only provision on Brazilian norms regarding involvement of other agencies in the EIA procedure. In practice, the environmental agency sends letters to other agencies informing about the EIA. However, as it was mentioned, agencies are generally overlooked and therefore do not participate in the procedure or participate at later stages.

Thus, based on the information provided in the environmental feasibility study, the project proponent should map stakeholders in order to identify important actors to participate in the scoping workshops. In this regard, research has demonstrated the efficiency of utilization of stakeholders' analysis (including mapping stakeholders) before the installation of a mining project in the state of Minas Gerais, in the southeastern region of Brazil. It generated a positive feeling on the project, since the company was making efforts to identify the potentially affected community and therefore open a channel of dialogue with them.[99]

The activity of mapping stakeholders should be regulated, establishing the area of influence as the baseline to identify them. This measure would avoid situations such as what happened in Volta Grande in which the EIA did not consider indigenous communities and FUNAI. The Brazilian environmental legal system should be uniform and adopt the area of influence as a common unit to consider impacts and stakeholders.

In sum, the utilization of identification of stakeholders' techniques would resolve a recurrent problem, which is the noninvolvement of important actors from the beginning of the EIA. In order to do so, EIA developers and the environmental agency must consider the area of influence of impacts as the geographical limit to be mapped. The area of influence of impacts is discussed next.

The Area of Influence of Impacts

As briefly mentioned in chapter 3, CONAMA Resolution No. 001/86 determines that the EIA must define the geographical area directly or indirectly affected by the impacts of the project (the project area of influence). Additionally, the area of influence should consider the watershed where the project is planned to be located (Art. 5, III). The establishment of such area assists the EIA developer to delimitate the breadth of environmental impacts diagnosis (impacts on physical, biotic, and socioeconomic environments).[100]

The area of influence assists in defining stakeholders who should participate in the EIA procedure, ranging from administrative unities such as municipalities, states, or other countries to communities, indigenous peoples, quilombolas, riverbank communities, and governmental agencies such as the national indigenous affairs, land reform, and health agencies.[101] Thus, it might define places where public hearings might happen.[102] Establishing narrow areas of influence might exclude participation of important stakeholders who are directly and/or indirectly affected by the project's impacts. Additionally, they will be excluded from mitigation, compensation, and monitoring measures.[103] Consequently, it might diminish the cost of the project with these measures.

In Brazil, the establishment of an area of influence is determined by the EIA developer based on the project characteristics and its impacts in compliance with the TOR. According to Machado, the definition of an area of influence is not under the project proponent, the EIA developer, or the environmental agency's discretion, but the possibility of occurrence of significant impacts determines the area of influence of any project. The CONAMA Resolution No. 001/86 requirement of considering the watershed where the project is planned to be located to establish the area of influence is an area of geographical reference that serves as a rigid parameter.[104]

Although the CONAMA Resolution requires EIAs to establish one area, which will be directly and indirectly affected by the project, in practice, Brazilian EIAs have different classifications and divisions of areas of influence.[105] For instance, practitioners divide them into physical, biotic, and socioeconomic. There are at least ten definitions and categorizations of areas of influence, such as areas directly affected by the project, areas of direct influence, areas of indirect influence, surrounding areas, and diffuse areas, among others.[106] The fact that there is no standard definition and classification causes confusion and harms the establishment of areas of influence since EISs filed at the environmental agency have different classifications and combinations of areas of influence.[107] Different EIAs adopt different methodologies and often EIA developers classify areas which will be directly affected as areas indirectly affected by the project. In this regard, researchers have reassured the necessity of regulating standards methodologies which are used in EIAs.[108]

Furthermore, the definition of an area of influence is related to the definition of direct and indirect impacts, which are neither defined in CONAMA Resolution No. 001/86. The area of influence should not be established as a rigid geographic limit defined at the beginning of the EIA procedure, as it is commonly seen in practice. Conversely, it should be part of the process and subject to public participation, consultation, and modification. According to Sanchez, the establishment of an area of influence should be a hypothesis to be confirmed. Only after the identification of impacts is it possible to define such area.[109] One example of this practice was the area of influence of the Belo Monte Dam. Initially, the area of influence of the project was demarcated. However, it changed in order to encompass a larger area.[110]

In this regard, in the EU EIA Directive the forecasting of significant environmental impacts will be determined taking into consideration the magnitude and spatial extent of the impact (e.g., geographical area and size of the population likely to be affected).[111] In other words, it is not preestablished an area of influence of the project, but based on the evaluation of impacts it is determined its spatial extent. Overall, public participation from the beginning of the EIA procedure would minimize the shortcomings mentioned previously. However, some other measures are also necessary to clarify and unify the establishment of areas of influence at the federal level, which will reflect at state and local levels.

First, environmental agencies must require EIAs compliant with CONAMA Resolution No. 001/86 and adopt only one area of influence. Relatedly, CONAMA should issue a guideline establishing methodologies capable of defining such area.[112] The fact that there are many different divisions, classifications, and definitions of the area geographically affected by the project creates at least two difficulties: it makes the EIA longer, more complicated, and it also opens the possibility of manipulating the area of influence since normally areas considered as indirect influence receive less attention and consequently fewer public hearings, less compensation and mitigation, and fewer monitoring measures.[113]

Moreover, the environmental feasibility studies must identify an area of study, which would be confirmed or changed throughout the EIA procedure. It would also be attached to the TOR. Public participation on the TOR should also be held, which would consider the area of study previously pointed in the feasibility studies, the population concerns, and their local knowledge.[114]

INCREASED PUBLIC PARTICIPATION IN BRAZILIAN EIAs AS OPPOSED TO REALITY (BUREAUCRACY AND LEGAL ACTIONS INVOLVING EIAs)

In responding to the long-standing argument that the Brazilian EIA procedure is already slow and involves many legal actions in overburdened courts, it is necessary to consider and reflect on the reasons why Brazilian EIAs are slow.

First, the poor quality of EIS, in conjunction with the fact that environmental agencies lack qualified personnel to review them, results in the filing of legal actions by *Ministério Público*, who questions environmental feasibility of proposed projects and lack of assessment of impacts in the EIS presented. A study written by Hoffman and presented to the Brazilian House of Deputies points to several reasons for the weaknesses of environmental permitting procedure in Brazil, such as EISs being too long, of poor quality, and focusing too much on the diagnosis of impacts; permits issued with an excessive number of conditionalities, which are frequently postponed in different phases of the procedure; and excessive bureaucracy. The author states that environmental analysts use bureaucracy as a defense to political pressure to approve certain projects.[115] Furthermore, in relation to excessive intervention of *Ministério Público*, Hoffman points out that this is a result of lack of transparency of environmental permitting procedures and decisions on the approval of projects are relative.[116] In sum, there is no standard to guide environmental decisions. All these weaknesses delay the procedure.

Several cases presented throughout this book are examples of this assertion. For instance, the Volta Grande Project, which did not properly assess cumulative impacts of mining activities with Belo Monte Dam. Furthermore, EIS of São Luiz do Tapajós Dam missed the assessment of impacts of the dam on threatened species in the area.

Second, proposed projects and their impacts are presented to affected communities at a late stage of the EIA procedure; therefore, there is a feeling that decisions are imposed on them instead of discussed at the early stages of the EIA. In this context, the concept of social license has been gaining great attention among private sectors because it has already been demonstrated how beneficial it is to have communities' acceptance in relation to a project. It improves the image of the company and diminishes protests, legal actions, and tentative to halt projects' implementation. Finally, in some situations, lack of adequate public participation also leads to questioning EIAs in court.

These are just some examples of why Brazilian EIAs are questioned through legal actions and end up in longer procedures. Eventually, the procedure with greater levels of public participation proposed here might increase the process by a few months, but it will decrease the amount of legal actions involving EIAs since it will increase transparency, social control, and accountability. Furthermore, Gaventa and Barret state that "increasing citizens' voice will make public institutions more responsive to citizens' needs and demands."[117] Susskind states, in relation to public dispute resolution, that "it is necessary to go slow to go fast."[118] This saying also applies in relation to the involvement of stakeholders in EIAs. To put it simply, doing things right and taking more time at initial stages, the procedure will flow with greater results and therefore faster at later ones.

In the case of Brazilian EIAs, massive projects have long EIAs because of legal actions filed against them, which also involves great costs. Therefore, having a longer EIA that leads to a project with greater community acceptance and involvement of key agencies and other stakeholders is better than having a procedure that is constantly being argued in court and being stopped by legal action. The measures proposed here might not avoid dissatisfaction of all stakeholders in some cases, but it will certainly decrease complaints and legal actions, if the procedure is well-conducted.

Benefits of social acceptance of projects will be noticed in the long term as it reflects on the public image and perception of enterprises, less tentative to halt projects, less prone to protests, strikes, and blockades. However, quantification of costs of social conflicts is still an area of research that needs to be developed further. For instance, a study conducted by Davis and Frank in which several employees from extractive industries were interviewed. Several interviewees pointed to the challenge of quantifying social conflicts in relation to the project. They compared employees' safety with community support issues. Therefore, they stated that "Zero incidents can be equated more easily with good safety performance than 'zero blockades' can be equated with positive community relations."[119] Although reputational risk is known as fear to any business market value it is also difficult to quantify it. There is some literature which points to the benefits of having social acceptance of projects. However, in terms of transforming these benefits into quantifiable results, this is still an area to develop. For instance, Davis and Franks mention that[120]

> Research by the Wharton School of Management into Canadian gold mining juniors suggests that the degree of stakeholder cooperation or conflict at a company's mines is a critical component in calculating the financial market valuation of the publicly traded parent firms. The research showed that 63 percent of the market capitalization of these companies can be linked to the quality of their stakeholder engagement—double the percentage linked to the value of the actual gold in the ground. Looking beyond share price, efforts to quantify impacts on reputation are notoriously difficult.

Therefore, there are obvious benefits to increasing public participation within Brazilian EIA procedures. From communities' point of view, it would increase their citizenship and trust in institutions and project proponents.[121] It would also allow them to have a voice in decisions which affect their lives directly. From project proponents' point of view, it would diminish reputational risks, which in turn reflects on financial gains, yet difficult to calculate. Furthermore, it would increase transparency and accountability in the EIA procedure.

CONCLUSION

This chapter demonstrated the great importance scoping represents for the effectiveness of the EIA procedure. Some countries' norms determine that scoping is an open and early process such as the United States. Others do not have much guidance in this phase, such as Brazil.

It was stressed that public participation would improve the entire EIA procedure when the public is involved in the scoping. The TOR guides EIA developers on what the EIA will cover. In Brazil, although there are several CONAMA resolutions establishing provisions on the development of the TOR, these provisions are related only to specific activities (oil and gas, transmission lines). Thus, there is no guidance on who should write the TOR for EIAs in general, and it is not open to public participation. This fact weakens the entire procedure because the public only participates once the EIS is already completed.

Considering the complexity of involving stakeholders who might be interested in the EIA procedure of certain projects, including governmental agencies, the following questions are raised: How should the TOR be elaborated? Who should participate in the development of the TOR and how?

This chapter proposes that TOR should be open to public discussion and consultation since it would allow participation at early stages of the EIA procedure, which provides more chances for the public to influence the decision-making in approving or rejecting projects.

While the draft of the TOR should always be open to public participation, massive projects (for instance, hydroelectric power plants, mining, fracking, among others) in sensitive areas (such as the Amazon rainforest) or projects to be located in areas where the cumulative impacts would cause significant environmental degradation would carry out scoping workshops with the affected population. To guarantee impartiality and meaningful communication the scoping workshops would be conducted by a neutral facilitator and supervised by an independent committee. A final version of the TOR must be posted at the agency's website and also made available to the communities. Having the possibility of a public hearing to discuss its final version.

To identify who should participate in the development of the TOR, those projects listed in the screening criteria would map stakeholders and present the results at the workshops for contributions. Stakeholders would be identified considering the area of influence of the project. In this regard, Brazil must provide technical guidance on how to define areas of influence as well as determine one area of study at the beginning of the process, having the possibility of extension.

Having a stricter procedure in terms of public participation at the commencement of the EIA utilizing techniques appropriate to different publics would allow meaningful participation in the entire process, since the EIA should be a process of education, consultation, and construction among all interested actors.

NOTES

1. Joe Weston, "EIA, Decision-making Theory and Screening and Scoping in UK Practice," *Journal of Environmental Planning & Management* 43, no. 2 (March 2000): 185, 188.
2. Ibid., 13.
3. Graham Wood et al., "EIA Scoping in England and Wales: Practitioners Approaches, Perspectives and Constraints," *Environmental Impact Assessment Review* 26, no. 3 (April 2006): 221, 222.
4. Ibid., 223.
5. Ibid.
6. The duration of an EIA depends on the country and its requirements.
7. Christopher Wood, *Environmental Impact Assessment: A Comparative Review*, 2nd ed. (London: Routledge, 2003).

8. Directive 2014/52/EU of the European Parliament and of the Council of 16 April 2014 Amending Directive 2011/92/EU on the Assessment of the Effects of Certain Public and Private Projects on the Environment, 2014 O.J. (L 124) 1.

9. See Wood et al., "EIA Scoping in England and Wales," 222; Weston, "EIA, Decision-making Theory," 188.

10. Weston, "EIA, Decision-making Theory," 200.

11. Ibid., 202.

12. Luis Enrique Sanchez, *Avaliação de Impacto Ambiental: Conceitos e Métodos* (Environmental Impact Assessment Concepts and Methodologies), 2nd ed. (São Paulo: Oficina de Textos 2013), chap. 6, pp. 14–30, E-BOOK, HYPERLINK "https://ebooks.ofitexto.com.br/product/avaliao-de-impacto-ambiental-2-ed" https://ebooks.ofitexto.com.br/product/avaliao-de-impacto-ambiental-2-ed.

13. G.E. Bealands and P.N. Duinker, *An Ecological Framework for Environmental Impact Assessment in Canada* (Halifax, NS: Institute for Resource and Environmental Studies, 1983), 66–67.

14. John Glasson, "Principles and Purposes of Standards and Thresholds in the EIA Process," in *Standards and Thresholds for Impact Assessment*, eds. Michael Schmidt et al. (Berlin: Springer, 2008), 9.

15. David P. Lawrence, *Impact Assessment Practical Solutions to Recurrent Problems and Contemporary Challenges*, 2nd ed. (Hoboken, NJ: Wiley, 2013), 227–258.

16. Ibid. For development of Social Impact Assessment (SIA), see Bianca Dendena and Stefano Corsi, "The Environmental and Social Impact Assessment: A Further Step Towards an Integrated Assessment," *Journal of Cleaner Production* 108 (December 2015): 969–972; Rabel J. Brudge et al., "Guidelines and Principles for Social Impact Assessment," *Environmental Impact Assessment Review* 15 (1995): 11. For Human Rights Impact Assessment, see Nora Gotzmann, "Human Rights and Impact Assessment," (Danish Institute for Human Rights, Matters of Concern Human Rights Research Paper Series 2014), http://www.humanrights.dk/publications/human-rights-impact-assessment.

17. Wood et al., "EIA Scoping in England and Wales," 225.

18. Ibid.

19. Ibid.

20. *See* Alexandra Polido and Thomas B. Ramos, "Towards Effective Scoping in Strategic Environmental Assessment," *Impact Assessment & Project Appraisal* 33, no. 3 (January 2015): 171–183.

21. Wood et al., "EIA Scoping in England and Wales," 224; Joseph Foti et al., *Voice and Choice: Opening the Door to Environmental Democracy* (World Resources Institute, 2008), 62.

22. Foti et al., *Voice and Choice*.

23. John Glasson et al., *Introduction to Environmental Impact Assessment*, 3rd ed. (London: Routledge, 2012).

24. *See* Lawrence, *Impact Assessment Practical Solutions*, 259.

25. Sanchez, *Avaliação de Impacto Ambiental: Conceitos e Métodos*, 14.

26. Ibid.

27. Ibid.

28. Foti et al., *Voice and Choice*, 47.

29. Wood et al., "EIA Scoping in England and Wales," 221–241.

30. Ibid., 224; Sanchez, *Avaliação de Impacto Ambiental: Conceitos e Métodos*, 16; Weston, "EIA, Decision-making Theory," 185.

31. Council on Environmental Quality (CEQ) Regulations for Implementing the Procedural Provisions of the National Environmental Policy Act, 40 C.F.R. pts. 1500-1508 (2015).

32. Wood et al., "EIA Scoping in England and Wales," 225.

33. *See* Impact Assessment Act S.C. 2019, c. 28, s. 1, art. 18 (1) (b) (1.2), https://laws.justice.gc.ca/PDF/I-2.75.pdf. *See* also Government of Canada, Impact Assessment Agency of Canada, Policy and Guidance, Impact Assessment Process Overview, Phase 1: Planning, (last accessed January 18, 2020) https://www.canada.ca/en/impact-assessment-agency/services/policy-guidance/impact-assessment-process-overview/phase1.html.

34. Wood et al., "EIA Scoping in England and Wales," 225.

35. Ibid., 226.

36. Ibid.

37. Ibid.

38. CONAMA Resolution No. 001/86, art. 5.

39. Ibid., art. 6.

40. An overview of the 2011 CONAMA Resolutions can be found in Aline Borges do Carmo, "A Dimensão Política do Atual Processo de Agilização Licenciamento Ambiental no Brasil" (The Political Dimension of the Current Process of Changing Environmental Licensing in Brazil to a Faster Procedure), *Revista Gestão e Políticas Públicas* 3 (December 2013): 285.

41. Portaria Interministerial No. 60, de 24 de Marco de 2015, Diário Oficial da União (D.O.U.) de 25.03.2015 (Braz.).

42. Portaria Interministerial No. 421, de 26 de Outubro de 2011, Diário Oficial da União (D.O.U.) de 28.10.2011 (Braz.).

43. Portaria Interministerial No. 422, de 22 de Outubro de 2011, Diário Oficial da União (D.O.U.) de 28.10.2011 (Braz.).

44. Portaria Interministerial No. 60/15.

45. Portaria Interministerial No. 421/11, art. 4, II.

46. Ibid.

47. Portaria Interministerial No. 422/11.

48. Ibid., art. 21, II.

49. Sanchez, *Avaliação de Impacto Ambiental: Conceitos e Métodos*, chap. 6, p. 29.

50. CEPRAM Resolução No. 2929, de 18 de Janeiro de 2002, Diário Oficial do Estado da Bahia (D.O.E.B.A.) de 18.01.2002 (Braz.).

51. Francisca Neta Andrade Assunção et al., "Participação Social na Avaliação de Impacto Ambiental: Lições da Experiência da Bahia" (Social Participation in Environmental Impact Assessment: Lessons from the Experience in Bahia), *Revista Franco Brasileira de Geografia* 10 (2010), https://confins.revues.org/6750.

52. *See* Sanchez, *Avaliação de Impacto Ambiental: Conceitos e Métodos*; Lei No. 1.356, de 3 de Outubro de 1988, Diário Oficial do Rio de Janeiro (D.O.E.R.J.) de 18.11.2008 (Braz.).

53. Antonio F.G. Beltrão, *Aspectos Jurídicos do Estudo de Impacto Ambiental (EIA)* (Legal Aspects of Environmental Impact Assessment [EIA]) (São Paulo: MP Editora 2008), 47.

54. *See* Jason Corburn, "Bringing Local Knowledge into Environmental Decision-making: Improving Urban Planning for Communities at Risk," *Journal of Planning Education & Resources* 22 (June 2003): 420. The author points to the definition of local knowledge and provides examples of cases where community knowledge improved projects implementation/operation.

55. Daniel Delong and Michel Fox, "Small Communities and Public Participation in the Cumulative Environmental Impact Assessment Process: The Case of the Little Bouctouche River, New Brunswick," *Journal of New Brunswick Studies* 1 (2015), http://w3.stu.ca/stu/sites/jnbs/en/issue_06-1.html.

56. Andréa Zhouri, "Justiça Ambiental, Diversidade Cultural e Accountability" (Environmental Justice, Cultural Diversity and Accountability), *Revista Brasileira de Ciências Sociais* 23 (October 2008): 97, 2012.

57. Beltrão, *Aspectos Jurídicos*, 50.

58. Ibid.

59. Ibid.

60. Environmental Resources Management, European Commission, *Guidance on EIA Scoping* (June 2001), http://ec.europa.eu/environment/archives/eia/eia-guidelines/g-scoping-full-text.pdf.

61. Ibid., 15.

62. Fernando Antonio Dorna Magalhães, "Licenciamento Ambiental. Atuação do TCU." (Environmental Licensing. The Role of the Audit Court) (PowerPoint presentation, Environmental Licensing: Reality and Perspectives, Brasília, November 5, 2015), http://www.mpf.mp.br/atuacao-tematica/ccr4/dados-da-atuacao/eventos/Seminarios/Ppt0000053.pdf.

63. The Cultural Palmares deals with issues related to the remaining quilombos communities (communities settled by runaway slaves).

64. Vanessa Alvarenga Caldeira, "Povos Indígenas, Sociedade Moderna e Projetos (Ditos) de Desenvolvimento Nacional" (Indigenous Peoples, Modern Society and (so called) Projects of National Development), *Tessituras: Revista de Antropologia e Arqueologia* 3, no. 1 (2015): 170, https://periodicos.ufpel.edu.br/ojs2/index.php/tessituras/article/view/5556/4129.

65. Portaria Interministerial No. 60/15.

66. Ibid.

67. Borges do Carmo, *A Dimensão Política do Atual Processo*, 290–91; Caldeira, "Povos Indígenas, Sociedade Moderna," 170.

68. Interministerial Ordinance No. 60/15, art. 3.

69. Ibid., Annex I.

70. Ibid.

71. Caldeira, "Povos Indígenas, Sociedade Moderna," 170.

72. Portaria Interministerial No. 60/15, art. 3, § 2, I.

73. Ricardo Verdum, "Hydroelectric Power Invades Brazilian Indigenous Lands," *Americas Program* (blog), September 8, 2015, http://www.cipamericas.org/archives/16168#.

74. Bom Dia Brasil, "Desmatamento Aumenta na Região da Usina de Belo Monte" (Deforestation Increases at the Belo Monte Dam Region), *G1*, October 27, 2015 8:53 a.m., http://g1.globo.com/bom-dia-brasil/noticia/2015/10/desmatamento-aumenta-na-regiao-da-usina-de-belo-monte-diz-instituto.html. The source discusses the consequences of the Belo Monte Dam to indigenous communities.

75. Alan Gilpin, *Environmental Impact Assessment: Cutting Edge for the Twenty-First Century* (Cambridge: Cambridge University Press, 1996), 20.

76. William A. Ross, "The Independent Environmental Watchdog: A Canadian Experiment in EIA Follow-up," in *Assessing Impact: Handbook of EIA and SEA Follow-up*, eds. Angus Morrison-Saunders and Jos Arts (London: Routledge: 2004), 178, 180.

77. Ismo Polonen et al., "The Effectiveness of the Finnish EIA System—What Works, What Doesn't, And What Could be Improved?," *Environmental Impact Assessment Review* 31, no. 2 (March 2011): 120, 123.

78. Zhouri, "Justiça Ambiental," 102.

79. Ross, "The Independent Environmental Watchdog," 178, 180.

80. W.E. Westman, *Ecology, Impact Assessment and Environmental Planning* (New York: Wiley, 1985). Glasson states that, "Public participation is likely to be greatest where public comments are most likely to influence decisions." Glasson et al., *Introduction to Environmental Impact Assessment*, 170.

81. Glasson et al., *Introduction to Environmental Impact Assessment*, 165.

82. Ibid., 160.

83. Beltrão, *Aspectos Jurídicos*.

84. Lawrence, *Impact Assessment Practical Solutions*, 265.

85. Ibid., 241, 244. The author discusses power imbalances.

86. Paulo Cesar Gonçalves Egler, "Improving the Environmental Impact Assessment Process in Brazil" (PhD diss., School of Environmental Sciences University of East Anglia, 1998), 63, http://busca.ibict.br/SearchBDTD/search.do?command=search&q=+instituicao_defesa:%22University%20of%20East%20Anglia%22.

87. Environmental Resources Management, *Guidance on EIA Scoping*.

88. Ibid., 23.

89. John Glasson et al., *Introduction to Environmental Impact Assessment*, 54.

90. *See* Luiz Enrique Sanchez, "Avaliação de impactos para licenciamento ambiental no Brasil: Modernização fundamentada no aprendizado." (Environmental Assessment for Environmental Licensing in Brazil: Updates Based on Lessons Learned), (PowerPoint presentation, Environmental Licensing: Reality and Perspectives, Brasília, November 5, 2015), http://www.mpf.mp.br/atuacao-tematica/ccr4/dados-da-atuacao/eventos/Seminarios/Ppt0000055.pdf.

91. "Public Participation Guide: Process Planning," United States Environmental Protection Agency (last accessed February 15, 2015), http://www2.epa.gov/international-cooperation/public-participation-guide-process-planning#step2.

92. International Finance Corporation, *Stakeholders Engagement: A Good Practice Handbook for Companies Doing Business in Emerging Markets* (Washington, DC: International Finance Corporation, 2007), 18, https://www.ifc.org/wps/wcm/connect/affbc005-2569-4e58-9962-280c483baa12/IFC_StakeholderEngagement.pdf?MOD=AJPERES&CVID=jkD13-p.

93. Ibid.

94. Ibid.

95. Ibid.

96. Ibid.

97. Glasson et al., Introduction to Environmental Impact Assessment.

98. 40 C.F.R. pt. 1501.6 (2015).

99. Lucas Silvestre de Carvalho, "Análise de Stakeholders com Stratégia de Relacionamento Externo: Estudo de Caso de uma Empresa de Mineração em Minas Gerais" (Stakeholders Analysis with External Relationship Strategy: Case Study of a Mining Company in Minas Gerais) (Master's thesis, Universidade Federal de Lavras, 2013), http://repositorio.ufla.br/jspui/handle/1/1016.

100. Willian Fonseca and Omar Yazbek Bittar, "Critérios para Delimitação de Áreas de Influência em Estudo de Impacto Ambiental" (Criteria for Delimitation of Area of Influence in Environmental Impact Assessments) (paper presented at First Brazilian Conference on Impact Assessment, October 15, 2012, https://www.scribd.com/document/353966779/CRITERIOS-PARA-DELIMITACAO-DE-AREAS-DE-INFLUENCIA-EM-ESTUDOS-DE-IMPACTO-AMBIENTAL.

101. Quilombolas are descendants of African slaves who maintain their religions and cultural traditions. "Fundação Cultural Palmares" (Palmares Cultural Foundation), Governo Federal de Brasil, (last accessed October 1, 2015), http://www.palmares.gov.br/?page_id=88&estado=RO. Ribeirinhos means riverbank dwellers. See also Procuradoria Geral da República, *A Área de Influência no EIA: Análise e Propostas* (The Area of Influence in Environmental Impact Assessments. Analysis and Proposals) (4CCR PGR, Nota Técnica 039/2007, October 17, 2007), http://www.mpf.mp.br/atuacao-tematica/ccr4/dados-da-atuacao/grupos-de-trabalho/gt-licenciamento/documentos-diversos/informa-tecnicas/NT%20039-07_Encaminha_trabalho_area_de_influencia.pdf/view.

102. Procuradoria Geral da República, *A Área de Influência no EIA*.

103. Ibid.

104. *See* Paulo Affonso Leme Machado, *Direito Ambiental Brasileiro* (Brazilian Environmental Law), 21st ed. (São Paulo: Malheiroso Editores, 2013).

105. Ministério Público da União (4a Câmara de Coordenação e Revisão—Meio Ambiente e Patrimônio Cultural), *Deficiências em Estudo de Impacto Ambiental* (Shortcomings of Environmental Impact Assessments) (2004).

106. Procuradoria Geral da República, *A Área de Influência no EIA*.

107. Ibid.

108. Sanchez states that there is great focus on the process and minimal attention on technical guidance. Sanchez, "Modernização fundamentada no aprendizado."

109. Sanchez, *Avaliação de Impacto Ambiental: Conceitos e Métodos*.

110. Leme Engenharia Ltda., *Estudo de Impacto Ambiental Belo Monte, Áreas de Influência* (Environmental Impact Statement of Belo Monte, Areas of Influence), (last accessed November 29, 2019), http://philip.inpa.gov.br/publ_livres/Dossie/BM/DocsOf/EIA-09/Vol%2005/AREAS%20DE%20INFLUENCIA/TEXTO/Cap%C3%ADtulo%206-%C3%81reas%20de%20Influ%C3%AAncia.pdf.

111. Directive 2014/52/EU, 2014 O.J. (L 124) 1.

112. Procuradoria Geral da República, *A Área de Influência no EIA*, 18.

113. Ibid.

114. Ibid.

115. Rose Mirian Hoffman, *Gargalos do Licenciamento Ambiental Federal no Brasil* (Shortcomings of the Federal Environmental Permitting in Brazil) (2015), http://bd.camara.gov.br/bd/handle/bdcamara/24039.

116. Ibid.

117. John Gaventa and Gregory Barrett, "So What Difference Does It Make? Mapping the Outcomes of Citizen Engagement" (Working paper No. 347, Institute of Development Studies, 2010), http://www.gsdrc.org/document-library/so-what-difference-does-it-make-mapping-the-outcomes-of-citizen-engagement/.

118. Lawrence Susskind, "Multi-party Public Policy Mediation: A Separate Breed," *Dispute Resolution Management* 4 (Fall 1997): 5.

119. Rachel Davis and Daniel M. Franks, *Costs of Company-Community Conflict in the Extractive Sector* (Cambridge, MA: Harvard Kennedy School, 2014), https://www.csrm.uq.edu.au/media/docs/603/Costs_of_Conflict_Davis-Franks.pdf. See Gaventa and Barrett, "So What Difference Does it Make?."

120. Davis and Franks, *Costs of Company-Community Conflict.*

121. Gaventa and Barrett, "So What Difference Does It Make?"

Chapter 6

The Environmental Impact Statement (EIS)

The EIS materializes the EIA process. In broad terms, it is the report that compiles data collected concerning the environmental baseline, the project's alternatives, the prediction of impacts, and mitigation measures, according to the previously agreed upon TOR. A well-prepared EIS guides the decision-making on the true consequences of the proposed project. For instance, uncertain and unavoidable adverse impacts, once clearly stated in the EIS, could lead to major modifications of the project or even its rejection by the authorization agency.[1]

Best practices indicate that a list of consulted stakeholders and how they were consulted, methods that have been used, difficulties that have been faced and limitations of the EIA should be part of the EIS.[2] Furthermore, a summary of the project and key issues at the cover of the EIS facilitates comprehension and enhances communications.[3]

The EIS of the Mount Hope Mining Project—for implementation in Eureka County, Nevada, United States—had the following abstract upfront:

> The Mount Hope Project is located on public land administered by the Bureau of Land Management and on private land controlled by Eureka Moly, LLC. The 80-year project would have an 18- to 24-month construction phase, 44 years of mining and ore processing, 30 years of reclamation, and five years of post-closure monitoring. Concurrent reclamation would not commence until after the first 15 years of the Project. The Mount Hope ore body contains approximately 966 million tons of molybdenite (molybdenum disulfide) ore that would produce approximately 1.1 billion pounds of recoverable molybdenum during the ore processing time frame. Approximately 1.7 billion tons of waste rock would be produced by the end of the 32-year mine life and approximately 1.0 billion tons of tailings would be produced by the end of the 44 years of ore processing.

> Optimal development of the molybdenum deposit, to meet the market conditions and maximize molybdenum production, would utilize an open pit mining method and would process the mined ore using a flotation and roasting process. The surface disturbance associated with the proposed activities totals 8,355 acres on both public and private lands.[4]

The nontechnical summary, as the name already denotes, is the synthesis of the EIS in accessible language to laypersons.[5] It is mainly through this document that the community will be informed about the proposed project. Generally, it is the basis for discussions and clarification of questions one might have.

Attention should be given to the inclusion of environmental justice, socio-economic dimension, and human rights impacts in the EIS, which might result in the implementation of the proposed project.[6] Moreover, an important matter concerning the development of the EIS is the impartiality of those who are in charge of this duty. Thus, the next section discusses different actors responsible for writing the EIS as well as solutions for preventing subjectivity in its development.

PARTIALITY AS OPPOSED TO TRANSPARENCY

Normally, project proponents write the EIS or hire consultants to write the document. The main issue related to this practice is that the project proponent has a direct interest in having its project approved. In the same fashion, the consultant chosen and hired directly by the project proponent is willing to have the project approved in order to have other contracts and keep the client satisfied.[7] Therefore, there is a high probability of biases in writing the EIS. Gilpin calls it the "ethical dilemma"[8]—where professionals face the difficulty of writing an EIS objectively or getting the project approved in the shortest period of time possible and therefore making their clients happy. It is common to see some EISs that exaggerate the benefits of proposed projects and hide or use "carefully chosen words" to describe the socio-environmental impacts.[9]

Independence, neutrality, and expertise are the most important characteristics EIS writers must have. It guarantees quality and transparency to the written assessment. Another current problem of EISs is the poor quality of these documents where impacts are superficially assessed or not assessed at all.[10] To guarantee impartiality and quality, countries have different systems. In the United States, applicants of private project permits are not allowed to write the EIS; they might prepare the initial Environmental Assessment (EA). However, the responsible governmental agency writes it or hires a consultant to write it at the expense of the project proponent.[11] Poland and the Czech Republic both adopted a list of licensed governmental consultancies[12] to ensure competence of professionals; however, this system also fosters corruption and discourages competitiveness.[13]

Independent EIA committees are also a practice in some countries, such as Finland, the Netherlands, Canada, and Poland. These committees review the EIS and sometimes act as channels of communications to the public.[14] However, most of the time the independent committees provide non-legally binding advice. The effectiveness of independent committees in assuring quality EISs has been recognized in Finland, where the Regional Center for Economic Development, Transport and the Environment is in charge of reviewing the EIS—from the scoping until the final document. It is a liaison on environmental matters. The strengths of this system are twofold: firstly, the Regional Center has expertise in revision of EISs and conducting the process as a whole; secondly, it is not responsible for issuing authorization permits. Therefore, it has a great level of independence. The Administrative Court has recognized its expertise and rarely contradicts the liaison authority.[15]

In Canada, the minister of the environment may, if he or she is of the opinion that it is in the public interest, refer the impact assessment of projects to a review panel, which will review the EIS developed by the proponent and ensure public participation in the EIA process, including holding public hearings.[16] The review panel will also prepare a report on the proposed EIS, which will indicate its conclusions and recommendations for the proposed project. The minister of the environment appoints the members of the review panel, who must be unbiased and free of conflicts of interest regarding the proposed project.[17] In the Netherlands, the Independent Netherlands Commission for Environmental Assessment (NCEA) reviews the quality of the EIS, potentially appointing experts and consultants to assist in this duty.

Voluntary initiatives also aim to improve the quality and independence of EISs. For instance, in the UK, the Institute of Environmental Management and Assessment (IEMA) created the "EIA quality mark." Consultants registered at IEMA must agree with the sustainability and ethical principles of the institution. Furthermore, EISs written by these consultants are independently reviewed by IEMA.[18]

In Brazil, federal, state, or sometimes local environmental agencies review the EIA. Under Article 7 of CONAMA Resolution No. 001/86, the EIA would be developed by a competent multidisciplinary team, which was not directly or indirectly dependent on the proponent of the project and would be technically responsible for the results presented. According to Article 7, the independence of EIA developers was highlighted. However, Article 7 was revoked by CONAMA Resolution No. 237/97, which was a setback in guaranteeing neutrality to the process. The current requirement is stated in

Article 11 of CONAMA Resolution No. 237/97, according to which, "[t]he environmental studies necessary as part of the environmental permitting procedure must be developed by legally qualified professionals at the expense of the entrepreneur."[19]

Therefore, considering the above provision, project proponents are entitled to realize the assessment and write the EIS. However, in general, project proponents outsource this duty to a consultancy because it requires various areas of expertise, which companies normally do not have among their employees.

Although consultants directly chosen and hired by project proponents might be partial in writing the EIS, they are legally responsible for the information they provide in the EIS. In this regard, Article 11 of CONAMA Resolution No. 237/97 states that "[t]he entrepreneur and the professionals who develop the environmental studies described in this article will be responsible for the information submitted by them, and will be subject to administrative, civil and criminal sanctions."[20]

Furthermore, Law No. 9.605 of December 2, 1998, prescribes reclusion from one to six years for those who present environmental studies that are partially or entirely false or misleading, including by omission.[21] The penalty might increase to one- or two-thirds of the final sanction fixed by the judge if there is a significant environmental impact as a result of the act or omission.[22] Although the fact of being criminally punished—even in case of omission—poses a serious threat to EIA developers, it does not improve the quality of the EIS, as has been seen in practice. Furthermore, it does not prevent EIA developers from being partial by exaggerating the benefits of the project. Therefore, an independent committee, which would participate throughout the elaboration of the TOR, would guarantee the impartiality and quality of the EISs.

The World Bank Report demonstrates that a major problem is that many Brazilian EIAs do not comply with the TOR.[23] One reason for this is the lack of social control over TORs, as the public is normally not consulted during its development there is not any reference to compare the EIS with its guidelines. For instance, when stakeholders consult the nontechnical summary of the EIS it is not known what was agreed in the TOR.

Another reason is the lack of personnel in environmental agencies available to review EISs. In the state of Pará, for example, SEMAS has only sixteen staff working on permitting procedures.[24] Additionally, some of them only work as part-time or temporary consultants. This is not enough personnel considering the dimensions of Pará and the complexity of the region; it is one of the Brazilian Amazonian states where there are several requests for authorization of projects in areas that face social and environmental challenges, such as the Volta Grande Project. Obviously, this weakness of environmental agencies is well-known by project proponents; thus, if some points of the TOR are not covered in the EIS, whether intentionally or not, the truth is that in many cases it may pass unnoticed by agencies responsible for reviewing EISs. Thus, having an independent committee would work as a "watchdog" to check whether the TOR is being followed or not.

BRAZILIAN EIS AND EIS NONTECHNICAL SUMMARY

Another main shortcoming of Brazilian EISs is that they are very long. For example, the EIS of the Belo Monte Dam was 20,000 pages long, in thirty-five volumes.[25] Similarly, the hydraulic power plant, São Luiz do Tapajós, was 15,000 pages long, in twenty-five volumes.[26] CONAMA Resolution No. 001/86 does not set page limits for EISs. Conversely, the CEQ regulations state that an EIS should have a maximum of 150 pages, and 300 in more complex cases. In the words of Barros, "The avalanche or excess of information might hide the information itself. Too much information does not necessarily guarantee the right to information, as this can possibly exceed the human capability to seize and cause rejection and disinterest."[27]

Even though some EISs are so extensive, it is not rare that legal actions are brought against project proponents due to lack of assessment regarding impacts on important social and environmental matters. The EIS of the hydraulic power plant of São Luiz do Tapajós is one example; although 300 professionals were involved in the EIA, the EIS (despite being very long) lacked information of impacts on threatened animal species who are restricted to portions of the Tapajós River, which led Greenpeace to develop a document entitled "Dams in the Tapajós River: A Critical Analysis of the EIA of the Hydraulic Power Plant São Luiz do Tapajós," written by independent experts in different environmental fields. Furthermore, the Federal *Ministério Público* recommended IBAMA not authorize the power plant, even though the government had previously announced that it would not give up implementing São Luiz do Tapajós Dam.[28] However, after protests from civil society and intervention of the *Ministério Público* in August 2016, IBAMA denied the authorization of São Luiz do Tapajós.[29] In sum, long EISs do not always mean quality or comprehensive studies. Lengthy EISs can cause confusion and lose focus on the main issues.

The instrument used for informing the public of the potential social and environmental impacts resulted from the proposed project is the EIS nontechnical Summary (in Portuguese *Relatório de Impacto Ambiental* or RIMA). Article 9 of CONAMA Resolution No. 001/86 states that the RIMA reflects the conclusions of the EIS and indicates a detailed list of items the summary must cover, including the objectives and justification of the project; a description of the project along with technological and locational alternatives; raw material which will be used; labor; generation of effluents; emissions; direct and indirect employment as a result of the project; description of the potential environmental impacts; considering the project, its alternatives, duration of impacts, methods and techniques adopted in the identification, quantification, and interpretation of impacts; the environmental diagnosis of the area of influence; the characterization of the future environmental quality of the area of influence; comparing the alternatives indicated, including the no-action alternative and the recommendation of the most favorable alternative.

Article 9 also states that the summary must describe the expected effects resulted from the mitigation measures proposed to minimize the negative impacts, also mentioning those impacts that are unavoidable and the level of environmental alteration expected as well as the impact-monitoring programs. According to Article 9 of CONAMA Resolution No. 001/86, the RIMA must be written in accessible language, which is not frequently seen in Brazilian practice. The next chapter discusses the barriers to access to information, including language.

CONCLUSION

The task of writing the EIS is normally carried out by consultants hired by the project proponents. This practice has been criticized by practitioners due to the fact that the consultant or the project proponents are partial, leading to bias.

In Brazil, although the law severely punishes those who omit or provide false information in EISs, bias in EISs has been a major problem, as mentioned by prosecutors, nongovernmental organizations, and the people.[30] Therefore, the adoption of an independent committee of experts could resolve this issue. The committee would be responsible for checking the quality of the EIS, checking if the EIS was written according to the previously agreed upon TOR, and ensuring that the nontechnical summary is written in accessible language. Moreover, the committee would issue its opinion on the approval or not of the EIS. The independent committee, however, would not be responsible for issuing permits in order to guarantee its impartiality in the process.

NOTES

1. John Glasson et al., *Introduction to Environmental Impact Assessment*, 3rd ed. (London: Routledge, 2012), 6.
2. Ibid.
3. Ibid.
4. U.S. Department of the Interior Bureau of Land Management, *Mount Hope Project Final Environmental Impact Statement Volume I of III*, October 2012, https:/

/eplanning.blm.gov/epl-front-office/projects/nepa/76912/102398/125246/2012_MT_HOPE_FEIS_VOL_1.pdf.

5. Ibid.

6. "Environmental justice is the fair treatment and meaningful involvement of all people regardless of race, color, national origin, or income, with respect to the development, implementation, and enforcement of environmental laws, regulations, and policies." See "Environmental Justice," United States Environmental Protection Agency, last accessed January 15, 2017, https://www.epa.gov/environmentaljustice.

7. Alan Gilpin, *Environmental Impact Assessment: Cutting Edge for the Twenty-First Century* (Cambridge: Cambridge University Press, 1996), 22.

8. Ibid.

9. Ibid.

10. Sabrina McCormick, "The Governance of Hydro-electric Dams in Brazil," *Journal of Latin American Studies* 39 (May 2007), 227, 240.

11. Environmental Law Institute, Report of the International Roundtable on Environmental Impact Assessment and Public Participation in Environmental Decision-Making (1992) (on file with author).

12. Ibid.

13. Martin Branis, "A System of Certified Environmental Impact Assessment Experts in the Czech Republic," *Environmental Impact Assessment Review* 14 (March 1994), 203–208.

14. Environmental Law Institute, *Report of the International Roundtable.*

15. See Ismo Polonen et al., "The Effectiveness of the Finnish EIA System—What Works, What Doesn't, And What Could be Improved?," *Environmental Impact Assessment Review* 31, no. 2 (March 2011): 123.

16. Impact Assessment Act S.C. 2019, c. 28, s. 1, art. 36 (1).

17. Ibid., art. 42 (d).

18. Jos Arts et al., "The Effectiveness of EIA as an Instrument for Environmental Governance: Reflecting on 25 years of EIA Practice in the Netherlands and the UK," *Journal of Environmental Assessment & Policy Management* 14 (December 2012), 25. See also "EIA Q-Mark," Institute of Environmental Management and Assessment (IEMA), last accessed October 22, 2019, http://www.iema.net/eia-quality-mark.

19. *See* Antonio F.G. Beltrão, *Aspectos Jurídicos do Estudo de Impacto Ambiental (EIA)* (Legal Aspects of Environmental Impact Assessment [EIA]) (São Paulo: MP Editora, 2008), 43.

20. Resolução CONAMA No. 237, de 19 de Dezembro de 1997, art. 11, Diário Oficial da União (D.O.U.) de 22.12.1997 (Braz.).

21. Lei No. 9.605, de 12 de Fevereiro de 1998, Diário Oficial da União (D.O.U.) de 12.02.1998 (Braz.).

22. Lei No. 9.605/98 altered by Lei No. 11.284/06. *See* Beltrão, *Aspectos Jurídicos*, 43; Lei No. 9.605, de 12 de Fevereiro de 1998, Diário Oficial da União (D.O.U.) de 12.02.1998 (Braz.); Lei No. 11.284, de 2 de Março de 2006, Diário Oficial da União (D.O.U.) de 03.03.2006 (Braz.).

23. World Bank, *Environmental Licensing for Hydroelectric Projects in Brazil: A Contribution to the Debate* (March 2008), http://siteresources.worldbank.org/EXTWAT/Resources/4602122-1214578930250/Summary_Report.pdf.

24. Interviews with anonymous individuals, groups, and public officials involved with the environmental permitting of the Volta Grande Project, in Altamira, Senador José Porfírio, and Belém, Pará State, Brazil, July 2–9, 2015. These interviews revealed some of the challenges faced by the Environmental Agency.

25. Eve Bratman and Cristiane Bená Dias, "Development Blind Spots and Environmental Impact Assessment: Tensions between Policy, Law and Practice in Brazil's Xingu River Basin," *Environmental Impact Assessment Review* 70 (May 2018).

26. Greenpeace, Barragens do Rio Tapajós: Uma Avaliação Crítica do Estudo e Relatório de Impacto Ambiental (EIA/RIMA) do Aproveitamento Hidrelétrico do Rio São Luiz do Tapajós (Dams of the Tapajós River: A Critical Analysis of the Environmental Impact Assessment and its Non-technical Summary of the Hydroelectric Use of the São Luiz do Tapajós River) (2015), http://greenpeace.org.br/tapajos/docs/analise-eia-rima.pdf.

27. Lucivaldo Barros Vasconcelos, "O Estado (In) Transparente: Limites do Direito a Informação Socioambiental no Brasil," (The Lack of Transparency of the State: Limits of Social and Environmental Information in Brazil) (PhD diss., Universidade de Brasília, 2008), 277–278, http://repositorio.unb.br/handle/10482/4167?mode=full.

28. Greenpeace, *Barragens do Rio Tapajós*.

29. IBAMA rejected the authorization in São Luiz do Tapajós because it would flood indigenous lands and would force their relocation, which was unconstitutional; in addition, the EIA/EIS presented many flaws. "Ibama Arquiva Licença Ambiental para Hidrelétrica São Luiz dos Tapajós" (Federal Environmental Agency—IBAMA—Nulls Environmental License Authorizing São Luiz dos Tapajós Dam), G1, August 5, 2010, 10:42 a.m., http://g1.globo.com/pa/santarem-regiao/noticia/2016/08/ibama-arquiva-licenca-ambiental-para-hidreletrica-sao-luiz-do-tapajos.html.

30. Ministério Público da União (4a Câmara de Coordenação e Revisão—Meio Ambiente e Patrimônio Cultural), *Deficiências em Estudo de Impacto Ambiental* (Shortcomings of Environmental Impact Assessments) (2004).

Chapter 7

Public Participation

Public participation has been a central concept in sustainable development.[1] It is well-recognized as an important role of access to information about environmental management and decision-making processes. As early as 1972, Principle 20 of the Stockholm Declaration requires "free flow of up-to-date scientific information and transfer of experience must be supported and assisted, to facilitate the solution of environmental problems."[2] Furthermore, Sands asserts that among the twenty-seven principles contained in the 1992 Rio Declaration on Environment and Development, four of them refer to forms of access to information and participation.[3] This fact reassures the importance of public information and participation in environmental matters.

Notably, Principle 10 of the 1992 Rio Declaration states three obligations to national governments:[4] (1) individual access to information concerning the environment that is held by public authorities, including information on hazardous materials and activities in their communities; (2) the opportunity to participate in the decision-making processes; and (3) effective access to judicial and administrative proceedings. In broad terms, the three pillars of Principle 10 can be defined as easy access for the public to environmental information from public authorities. The information should be clear as well to nonexperts. Otherwise it becomes, ultimately, inaccessible.[5] Access to justice is the capacity of individuals or groups to seek remedies when the access to information and public participation is denied or is not clearly provided.[6]

Additionally, Chapter 40 of Agenda 21 also states that in sustainable development, "everyone is a user and provider of information."[7] Therefore, the "need of information arises at all levels from that of senior decision makers at the national and international levels to the grass-roots and individual levels."[8] Following the Rio Declaration, in 1998, the European Commission signed the Aarhus Convention on Access to Information, Public Participation and Access to Justice in Environmental Matters, which entered into force in 2001.[9] The convention represents progress toward public participation in environmental decision-making.

On March 4, 2018, in Costa Rica, the United Nations adopted the Regional Agreement on Access to Information, Public Participation and Justice in Environmental Matters in Latin American and the Caribbean: the Escazú Agreement. In September 2018, fourteen countries signed the agreement, including Brazil.[10] Implementation of a regional convention on public participation in environmental issues will reinforce the need to create effective mechanisms of public participation in a region facing socio-environmental conflicts due to implementation of mega projects.[11]

Wood asserts that "EIA is not EIA without consultation and participation"[12]—it is a fundamental part of the procedure. While different definitions on public participation have been advanced, for the purpose of participation in EIA, this book adopts Aber's concept where participation refers to "increasing citizen control over the state and improving the capacity of ordinary people to understand and decide about issues affecting their lives."[13]

Considering the classification of participatory designs, public participation in EIAs could be classified as power sharing, where negotiations and deliberations are mandatory.[14] However, this is not the case of many countries where participation is not meaningful. For instance, lack of transparency and access to information impairs participation. Additionally, participation techniques do not recognize special characteristics of different groups, such as indigenous communities, illiterate individuals, communities who live in distant areas, or others who, generally, are the most affected by development.[15] Normally, or at least in developing countries, there is an imbalance of powers in the participatory arena where those who hold economic powers have advantages in relation to others least favorable considering the traditional economic models.[16] Thus, it is necessary to develop methods capable of informing, engaging, consulting, collaborating, and educating the public.

According to Milaré, the EIA is subject to three conditions: (1) transparency, (2) consultation to stakeholders, and (3) the grounded decision on approving or rejecting the project by the public administration. This section discusses the first two conditions since transparency is a prerequisite for meaningful public participation and consultation.[17] EIA decisions are discussed in chapter 8.

TRANSPARENCY

Fox states that transparency is a "safeguard mechanism because it protects the public from abuse of powers."[18] In fact, transparency ensures social control over administrative decisions, exercised by both the public and governmental agencies.[19] Therefore, transparency in EIA means disclosure of documents and information to public consultation, appropriate communication of availability of EIS and its nontechnical summary, and grounded and accessible decisions demonstrating how public input was considered.

In several countries, transparency in EIAs is limited by industrial and trade secrets.[20] In Norway, the EIA regulation states that the agency can determine that certain parts of the EIA are confidential. Additionally, Article 10 of the EU EIA Directive states that, "the provisions of this Directive shall not affect the obligation on the competent authorities to respect the limitations imposed by national laws, regulations and administrative provisions, and accepted legal practices with regard to commercial and industrial confidentiality, including intellectual property, and the safeguarding of the public interest."[21]

In Brazil, Article 225 of the Federal Constitution expressly determines that EIAs shall be made public.[22] Furthermore, Article 5(XXXIII) states that

> all persons have the right to receive, from public agencies, information of private interest to such persons, or of collective or general interest, which shall be provided within the period established by law, subject to liability, *except for the information whose secrecy is essential to the security of society and of the State.* (emphasis added)[23]

Notably, the constitution makes one exception to people's right to receive information from public agencies, which is, when the disclosure of such information threatens the security of society and of the state.[24] However, Article 18, § 3 of Federal Decree No. 88351/83 states that "[r]especting industrial secrets, when expressly required by the proponent of the project, the EIS non-technical Summary Report will be accessible to the public."[25] CONAMA Resolution No. 001/86 repeats the same obligation of the aforementioned Federal Decree.

Machado says that the limitation of trade and industrial secrets should be interpreted in light of the Federal Constitution, which states that the EIS shall be public. Furthermore, the determination in suppressing certain parts of the EIS due to industrial secrets is subject to judicial revision, where the judiciary might reverse the administrative decision of nondisclosure of documents due to industrial secrets.[26]

According to Article 5(XXIX) of the constitution, "the law shall ensure the authors of industrial inventions of a temporary privilege for their use, as well as protection of industrial creations, property of trademarks, names of companies and other distinctive signs, viewing the social interest and the technological and economic development of the country."[27] In sum, industrial and trade secrets are not absolute rights. They are subject to the checks and balances in the interest of society and the development of Brazil.[28] Thus, whether the information should be omitted on the EIS or not is under the discretionary power of the agency.

ACCESS TO INFORMATION

Although some scholars still argue over whether access to information is an emerging or a fundamental right,[29] both the IACHR and the European Court of Human Rights have issued decisions recognizing the right to access information held by governments as a human right.[30] Additionally, access to information is explicitly stated in Article 13 of the American Convention on Human Rights, which reads, "Everyone has the right to freedom of thought and expression. This right includes freedom to seek, receive, and impart information and ideas of all kinds, regardless of frontiers, either orally, in writing, in print, in the form of art, or through any other medium of one's choice."[31] Therefore, governmental documents and decisions must be available to public consultation.

As far as the EIA is concerned, Wasserman points out that countries have different policies on what would be enough in terms of the disclosure of documents.[32] In some jurisdictions the entire EIS is available for public consultation; in others, only the nontechnical summary is disclosed. Additionally, in some countries the public is allowed to read the EIS but cannot make copies.[33]

The World Bank Environmental and Social Standard 10 states that "The borrower will disclose project information to allow stakeholders to understand the risks and impacts of the project, and potential opportunities. The borrower will provide stakeholders with access to . . . information, as early as possible, and before the Bank proceeds to project appraisal and in a timeframe that enables meaningful consultation."[34] This is a good provision, although many countries fall short on providing true access to information.

To ensure that the public is able to participate effectively in the EIA procedure, some actions should be taken: (1) make the EIS and its nontechnical summary available in places of public access (disclosure of documents), (2) acknowledge to the public that the EIS and its nontechnical summary is ready for consultation through different and effective channels of communication, (3) provide enough time for public consultation, (4) make communication (written and oral) available in language accessible to laypersons.[35]

Acknowledge to the Public that the EIS and Its Nontechnical Summary Is Ready for Consultation Through Different and Effective Channels of Communication

The Brazilian Federal Law No. 10650 of April 16, 2003, regulates public access to data and information held by agencies of the National Environmental

System (in Portuguese Sistema Nacional de Meio Ambiente or SISNAMA).[36] Under Article 4 information regarding the request of an EIA as well as its approval or rejection must be published in the Official Gazette and such information must also be available in places of easy public access.[37]

Article 10 of Federal Law No. 6938/81 (the Environmental Policy Act) states that the environmental permitting requests, and the issuance and renewal of environmental permits, will be published in the Official Gazette as well as in regional periodicals or local newspapers of great circulation, or in the environmental agency's electronic means of communication.[38] In this context, CONAMA Resolution No. 006/86 contains models of publications to inform the public on the necessity of developing an EIA or not for the specific project requiring an environmental permit.[39] Machado highlights the great importance of these provisions stating that the population has the right to know whether certain project requires an EIA or not.[40] Thus, the public can challenge the process at early stages. In fact, this should be the intention of communicating the public. However, it is hard to challenge a procedure where the public did not participate from the beginning (for instance, during the drafting of the TOR).

The announcement of an EIA procedure in official newspapers, such as official gazettes, does not reach a great portion of the population.[41] According to Beltrão, a much simpler measure than publishing in an official gazette would be posting on the environmental agency's website, which would be more efficient in informing the public.[42] Along these lines, two web-based tools—the NEPAssist and the EIA tracking system—are being implemented in Central America and the Dominican Republic in cooperation with the U.S. Environmental Protection Agency. These tools provide all EIA documents from the very beginning stages: from the application to the environmental permit and to audit reports. It also utilizes the geographical information system to visualize the project and its characteristics, such as boundaries, which assists in enforcing commitments made on the EIS.[43]

These are great solutions. Social media and websites can reach parts of the population that would be interested in participating in the EIA procedure. However, many do not have access to the internet, are illiterate, or speak other languages. Thus, the utilization of other means of communication would be more effective, in rural areas for example—such as radio and TV announcements on EIS and nontechnical summary availability for consultation. Pamphlets and outdoor advertisements would also be alternatives. These means of communication could be used in conjunction with social media and websites to reach all those who would be interested in participating in the EIA. While some private sector companies have already been carrying out these practices, they are not regulated or guided by any Brazilian governmental document.

Disclosure of EIS and Its Nontechnical Summary

Article 11 of CONAMA Resolution No. 001/86 determines that the EIS nontechnical summary (in Portuguese *Relatório de Impacto Ambiental* [RIMA]) will be available for public consultation, requiring copies at documentation centers or environmental agencies' libraries, including during the period of technical analysis. The CONAMA Resolution does not expressly require that the RIMA be available in the municipalities where the project will be located. According to Machado, however, copies of the EIS and RIMA should be available at municipalities since this practice is under the objective of the EIA procedure.[44]

In the same line, CONAMA Resolution No. 006/87, which regulates the environmental permitting procedure of power plants, only requires that RIMA will be available for public consultation while the EIS (meaning the complete and detailed technical study) should be analyzed by the authorizing and competent agency. Federal Decree No. 88351/83, which regulates the National Environmental Policy Act, contains similar provisions stating that RIMA will be available for public consultation.[45] However, Machado states that the objective of the Decree, although not expressly stated, is that both EIS and RIMA will be disclosed for public consultation.[46] Machado argues that there should not be two different assessments—one for the public and one for the agency. Instead both studies should be equally available to all.[47]

Only providing RIMA for public consultation is unconstitutional because the Federal Constitution states that the EIS must be made public.[48] In fact, an executive resolution such as CONAMA's cannot restrict public access to any information and documents related to the EIA.[49] Although there are legislations suggesting that only the RIMA would be available for public consultation, environmental agencies often make both EIS and RIMA available prior to public hearings—a process discussed later in this chapter.

The project proponent must present five hard copies of the RIMA to the environmental agency.[50] Environmental agencies commonly determine that the RIMA is available for consultation at its library. However, the community that will be directly affected is often located miles away from the agency. Consequently, communities do not have easy access to the EIS and RIMA. As Barros notes, the fact that access to information is written in a law or regulation does not mean that it is effectively guaranteed to the people.[51]

Governmental agencies that express interest or have direct relationships with the project will receive copies of the RIMA.[52] The environmental agency identifies these interested agencies. Specifically, the Environmental Ministry Guide on Environmental Permitting Procedure mentions the involvement of ten agencies—such as the National Department of Mineral Production, the Palmares Foundation, the Indigenous affairs agency, and others—referencing legislation that governs the procedure in relation to these agencies. Interestingly, the guide does not mention the specific involvement of the Land Reform Agency. No official list of stakeholders exists nor is there a requirement to carry out stakeholder analysis. In this sense, at early stages of the EIA, a mapping of stakeholders should be carried out to identify who should participate in the EIA and define the agencies that should receive the EIS and the RIMA.

The Language of the Nontechnical Summary

Article 9 of CONAMA Resolution No. 001/86 orients EIA developers on what RIMA must include—specifically, it must be presented in an objective and adequate manner for comprehension. The information must be translated into accessible language—illustrated with maps, charts, graphs, tables, and other techniques of visual communication—to facilitate an understanding of the advantages and disadvantages of the project as well as all the environmental consequences of its implementation. Although the objective of Article 9 is clear, in practice, EIA developers do not comply with this obligation. The message contained in RIMAs is often transmitted to everyone equally—as if everyone possessed identical knowledge and comprehension capacity for somewhat technical concepts.[53]

Language is one of the most powerful tools to inform the public about the project and its impacts. In this sense, Barros states that information only achieves its function if it is transmitted in a way that can be understood and assimilated by receptors. Therefore, the fact that EISs and its summary utilize technical language results in weak communication and represents a barrier to accessing information because laypersons cannot fully comprehend the content of documents written by (and for) professional experts in engineering, environmental sciences, chemistries, among others.[54]

In remote areas, where many major projects are implemented in Brazil, there is a high level of illiteracy as well as different dialects and languages. Instituto Socioambiental has identified 247 indigenous ethnicities and 154 dialects and indigenous languages spoken in Brazil.[55] Thus, the use of maps and photos might help comprehension of the projects and its potential impacts to diverse communities. The scoping workshops carried out prior to the EIS are of great importance in these cases.

Independent committees, proposed in chapter 8, can play a role in assuring that the language of the RIMA is appropriate to the recipients. In other words, written in accessible language, containing maps and other forms of visual communication, and shared in the languages spoken by affected communities.

PUBLIC CONSULTATION

While "public participation" is broadly referenced in this book, important differences exist between public consultation, participation, and negotiation. Consultation is a two-way communication among stakeholders, project proponents, and the authorizing agencies; it aims to inform the public and collect their opinion on the proposed project and its consequences. Participation is engagement of stakeholders in the proposed project and it aims to obtain consensus on the proposal. If disagreement over certain matters occurs, negotiation among stakeholders aims to settle conflicts on the proposed project. These mechanisms are defined as forms of public involvement.[56]

The United Nations University, in partnership with other institutions, developed the EIA Course, establishing common principles recognized as key for public involvement:[57]

- inclusive—covers all stakeholders;
- open and transparent—steps and activities are understood;
- relevant—focused on the issues that matter;
- fair—conducted impartially and without bias toward any stakeholder;
- responsive—to stakeholder requirements and inputs; and
- credible—builds confidence and trust.

In general, EIA procedures do not have real forms of public involvement.[58] For instance, regarding the Brazilian practice, Sanchez states that participation in the country happens in a "reactive model, in which the project proponent informs, and the community reacts."[59] Thus, there is no common construction of ideas or a two-way flow of communication. The following subsections focuses on specific forms of public consultation: written comments and public hearing. They analyze how these phases are carried out in Brazil in a comparative fashion, then explain how the measures proposed throughout this book would improve public participation in Brazil.

Written Comments

EIA procedures throughout the world have some form of public consultation. Normally, countries adopt a written comments period and sometimes a public hearing is held.[60] The phase of EIA when public comments happen varies. For instance, Canada's provinces have different rules. In Alberta, under the auspices of the Environment and Sustainable Resource Development, the public can provide comments only on the terms of reference; during other phases of EIA there is no such opportunity.[61] However, in Nova Scotia, the public has thirty days to comment on the registration of the project in the authorizing agency—even before the TOR.[62] Conversely, in the United States, the CEQ Regulations require the agency to invite comments on the EIS draft and again before preparing a final EIS from the public, other federal agencies, and tribal communities. And, before taking a final decision, if the agency thinks it is necessary.[63]

The United Nations University states that it is good practice to open a period of written comments during the EIA procedure. However, this is the minimum step of public consultation broadly practiced in EIA procedures around the globe.[64] The following are the important points to be considered: (1) whether written comments are effective, (2) how public comments are incorporated into the EIA review, (3) whether and how project proponents and authorizing agencies respond to public comments and concerns, (4) whether enough time is provided for the public to express views and/or clarify any questions in the form of written comments.

It is inarguably important to have written comments in EIA procedures and, although it might not reach all who will be potentially affected by projects, such as indigenous peoples who speak other languages, illiterate people, and others, the relevance of this phase should not be ignored. For instance, members of academia and experts can express, in a written format, their analysis and opinion on the proposed project. Such expression of opinion should be attached to the EIA process and best practice determines that stakeholders' comments should be appropriately addressed.

In this regard, the United States regulates the written comments period reasonably well. Section 1503.3 of the CEQ Regulations states that comments should be as specific as possible.[65] The lead agency must consider the comments made by modifying the project, consider other alternatives that were not included or alternatives that were not seriously assessed, supplement or modify its analysis, and make corrections or explain why the comments were not addressed.[66]

Similarly, in Ontario, Canada, a Code of Practice orients the ministry that approves the project to evaluate how public comments were considered and included in the TOR and the EIS. Non-consideration of comments must be explained.[67] In countries such as the United States and Australia, project proponents must provide written responses to public comments. Conversely, in Germany, the Netherlands, Portugal, and Brazil, no such obligation exists.[68] CONAMA Resolution No. 001/86 has a simple provision on written comments, stating that once the EIS and RIMA (nontechnical summary) are filed, the environmental agency will establish the period to collect written comments from both the public and other environmental agencies.

Machado argues that CONAMA Resolution No. 001/86 should have established a minimum period of time for gathering written comments of no less than thirty days.[69] The author also mentions that in other countries, the EIA legislation states the period for written comments. For instance, in Italy, the written comments period lasts thirty days and, in Chile, for sixty days.[70]

IBAMA Executive Order No. 184/08 establishes deadlines for federal environmental permitting procedures. It was amended in 2011 by Executive Order No. 14. The latter establishes that

> Art. 10 § 3° Agencies and federal entities involved in the construction of the TOR will be consulted according to deadline and procedures established in appropriate norms. § 4° State environmental agencies involved in the draft of the TOR shall intervene, in 15 days, in the necessary studies to the assessment of the project, its impacts and mitigation measures . . . Art. 21-A. State environmental agencies involved shall have thirty days to intervene on the project, its impacts and mitigation measures which be counted from the acknowledgement of the EIS. § 1° Lack of expression of opinion will be understood as the state environmental agency's agreement to the EIS presented.[71]

These provisions refer only to intervention of other entities or state environmental agencies. In relation to public written comments, it is up to the environmental agency (IBAMA in the case of federal procedures) to decide the deadline to collect written comments. In Brazil, whether the state environmental agency issues an opinion on the EIS within the deadline established indicates whether the state environmental agency agrees with the EIS presented. In other words, silence suggests agreement.

In this context, section 1503.2 of the CEQ Regulations has a better provision, which states that even if the intervenient agency is satisfied with the EIS presented it should reply that it has no comment.[72] The duty to comment is appropriate and demonstrates that the agency acknowledges the EIS—even if it did not have comments.

In Brazil, there is no obligation to respond or consider comments made by the public. Another shortcoming of the Brazilian practice and normative gap is the fact that there is only one opportunity for written comments on the EIS proposed, which is after the EIS is filed at the environmental agency. Thus, if the environmental or other governmental agencies request the project proponent to supplement the EIS or clarify some specific issues, such complementation and clarifications will not be subject to public consultation since the written comment period already happened. This was one of the findings of the *Ministério Público*'s report on Deficiencies of the Brazilian EIA procedure.[73] The report suggests that, in case of complementation of the EIS, there should be another phase of public consultation when the public is not aware of new information provided by the proponent of the project, which might be relevant for those who would live with the impacts of the project.[74]

In this sense, the U.S. NEPA is more appropriate because it provides a two-step EIS development: the draft EIS and the final EIS. Both may be subject to public comments.[75]

Incorporating the measures proposed in previous chapters—specifically, developing the TOR by listening to affected communities—would also assist the public in making substantive comments on the draft EIS. Additionally, further complementation should be available for public consultation in the public hearing. Certain tools would improve the written comments period. For instance, the development of a blog and website for this purpose of comments would improve engagement. Furthermore, an innovative technology, the Talking Book, would be a useful tool for distant communities since it is a simple audio computer device that can be heard anywhere. It is designed for users who have low literacy skills. The Talking Book disseminates messages and allows the user to record their feedback about certain topics.[76] Additionally, a mailing list could inform the public on how the comments were considered in the final decision.[77] These are good institutional practices that could be followed in Brazil. The independent committee proposed in this book could administrate the blog or website, collect people's messages from the Talking Book, and send emails to the mailing list.

Public Hearing

Public hearing is a method of public consultation. It is a formal space, usually with rules preestablished, where stakeholders can orally offer comments and clarify their questions on the proposed project.[78] Public hearings are important means of communication among communities, project proponent, and governmental agencies. They are tools of democracy in which everyone can participate and provide important information regarding the local environment, which can be used by EIA developers to improve the impact assessment documents (TOR and EIS).[79]

Additionally, hearings are important to present the final version of the TOR and to acknowledge stakeholders on the findings of the EIA in order to compare them with the reality of the area where the project is planned to be installed.[80] Although they are very important, public hearings should not be the only opportunity for the public to acknowledge the project's impacts and provide their views. As mentioned in previous chapters, public participation is a cyclical process and should happen from the very beginning of the EIA.

Some countries have thresholds to identify which project's EIA should encompass a public hearing before a decision is taken. For instance, in Ghana, projects that demand "relocation/resettlement of communities, strong public concern about environmental impacts, proximity to settlements, in the case of mines, diversion of railway lines, streams, power lines" require public hearings.[81] In the United States, public hearings are required when there is substantial environmental controversy over a project, when a hearing is required by another agency located in the same jurisdiction where the project will be located, or when it is necessary to gather information from the public.[82] In Norway, the public hearing takes place whenever the project might cause significant environmental pollution.[83] In Canada, the review panel identifies the necessity of holding a hearing. In Colombia, a public hearing is held if it is requested by 100 citizens or three nonprofit organizations, the governor, the environmental minister, or other environmental authorities.

Public hearings are costly, and time consuming. During hearings it is crucial to establish clear communications if communication is not appropriate, it leads to misunderstanding, which may cause communities' negative reaction toward the project or strikes.[84] To avoid or minimize conflicts, or to direct the discussions during the public hearings toward communities' concerns, prehearing meetings with stakeholders are considered best practice.[85] If the public hearing is the first opportunity for public interaction, by the time it happens it is already too late to influence decision-making since significant time and money will have already been spent on the EIA, and it is unlikely that governments will reject a project at this stage.[86]

Practice has demonstrated some flaws in the realization of hearings. In this regard, Westman lists methods of public participation according to their effectiveness, public inquiry (or hearing) being one of the least effective methods capable of affecting decision-making, followed only by litigation and protests.[87] Communication does not always flow well in public hearings due to the utilization of technical language and lack of prior stakeholder engagement, which consequently leads to confrontations instead of dialogue and discussions regarding the project and its potential impacts.

Therefore, to guarantee the effectiveness of public hearings, special attention should be given to its preparation and realization, such as providing appropriate time for the public to prepare for the hearing by making the EIS and its nontechnical summary available, having clear regulations on whether the hearing will be held, implementing prehearing meetings, selecting appropriate place and time that facilitates participation of stakeholders, using means of communication that reach participants (nontechnical language, indigenous languages, maps, images, photos, among others),[88] conducting the hearings by an unbiased team or moderator, and using mediation techniques.

Certain situations might create an intimidating environment, such as setting the room, so the proponent of the project and the environmental agency are sitting on an elevated stage. Creighton states that "the ideal arrangement for interaction or consensus forming/negotiation is a circle." A circular arrangement establishes eye contact among all participants. It also puts everyone in equal status, removing any sense of there being a "head of the table." Additionally, the presence of highly armed military force in the room and the occurrence of hearings in large "fancy" places instead of in places that are familiar to the affected communities such as schools, community associations, and churches can create an intimidating environment.[89]

Procedurally speaking, in some jurisdictions, the public hearing happens after the completion of the EIS and it is conducted by an independent panel—such as in Canada and Ghana.[90] In the latter, the panelists will indicate a chairperson for the hearing who is not a resident in the area of influence of the project. In other jurisdictions, the public hearing is held by the EIA reviewing agency, such as in New Zealand and Brazil, or by the governmental agency responsible for the project, such as in the Netherlands.[91]

Public Hearing in Brazilian EIAs

In Brazil, public hearings were established in the last part of CONAMA Resolution No. 001/86 and later regulated by CONAMA Resolution No. 009/87.[92] Although they have important roles in Brazilian EIA procedures, their realization has some weaknesses.

During interviews carried out by the author in the state of Pará concerning the case study analyzed in chapter 3, one member of a civil society organization noted that "The way participation happens in public hearings must change. There is no deliberation in these meetings."[93] The public prosecutor affirmed that during the public hearing regarding the EIA procedure of the Volta Grande Project, it felt as though the decision had already been made and the hearing was just happening to comply with the resolution. Milaré reiterates that "Brazilian democratic institutions are still fragile" referring to the effectiveness of public hearings in Brazilian EIA procedures.[94] Institutional practice in Brazilian EIA public hearings should change. It is necessary to adopt a preestablished methodology and techniques, where stakeholders truly communicate among themselves.

Article 1 of CONAMA Resolution No. 009/87 states that the objective of the public hearing is to expose to stakeholders the content of the project and its nontechnical summary, to clarify questions, and to gather critiques and suggestions. However, an equivocation exists in Article 1 of the CONAMA Resolution. The last part states that the public hearing will discuss the nontechnical summary, but, in fact, both EIS and the nontechnical summary must be available for discussion[95] as the Federal Constitution requires that the EIS be public.

According to Article 2 of CONAMA Resolution No. 009/87, the environmental agency has discretionary powers to request a public hearing if it finds one necessary. In other words, if the agency does not request it, the hearing might not happen, unless requested by a civil entity, the *Ministério Público*, or at least fifty citizens. If these actors request the public hearing and it does not occur, any environmental permit issued will be null. While there is a possibility of citizens requesting a public hearing, which then becomes obligatory, according to the resolution, there is also a possibility of no hearing occurring.

Benjamin states that the public hearing in the Brazilian EIA procedure materializes two important principles: public participation and the access to information principles.[96] As such, its occurrence should be obligatory, at least for massive projects, since—in the current Brazilian EIA procedure—the hearing is the only moment when the public participates. Furthermore, considering the methods utilized to inform the public about an EIA (publication in official gazette and newspapers), which might not reach all affected communities, it is not difficult to miss the deadline to request a public hearing.

The constitutions of some Brazilian states make a public hearing obligatory, such as in Goiás, Maranhão, Mato Grosso, Mato Grosso do Sul, Pernambuco, and São Paulo states.[97] Having obligatory public hearings in the EIA procedure is wise; furthermore, it fulfills the goal of the EIA procedure, which is to access all social and environmental impacts of a project. Thus, to access all possible impacts it is necessary to listen and consider the opinion of those who will be potentially affected by the project.

Once the environmental agency receives the EIS, it is published in the local newspaper starting a minimum period of forty-five days for a civil entity, *Ministério Público*, or fifty citizens to request a public hearing. Notably, CONAMA Resolution No. 009/87 only states that the EIS shall be published in a "local newspaper," which, if interpreted literally, would not achieve the goal of informing all stakeholders of the public hearing since the "local newspaper" would be understood as the municipality where the project would be located and not the area of influence or the hydrographic watershed, which typically encompasses more than one municipality.[98]

Moreover, the resolution only sets the period to request the public hearing; it does not set a minimum period of time to prepare for the hearing. In other words, the environmental agency could hold the hearing right at the end of the forty-five days, which could compromise the efficiency of the hearing since communities would not have adequate time to consult the EIS. For instance, if an organization requests a public hearing on the last day of the period (Day 45), the public hearing could happen within a week, within ten days, or any time after the end of the period, which may not provide enough time for the public to read the EIS and prepare for the hearing.[99] Legislation in the Rio Grande do Sul state in Brazil accounted for this issue, requiring the public hearing to happen thirty days or more after the announcement of the public hearing.[100]

At the end of the forty-five day period, if a public hearing is requested, the environmental agency will inform those who requested it through registered letters and publication in local media. Article 2, Paragraphs 4 and 5 of CONAMA Resolution No. 009/87 require the public hearings to occur in places accessible to interested parties. Moreover, considering the geographical localization of those who requested the public hearing and the complexity of the project, it is possible to have more than one public hearing on the same project.[101]

Therefore, the resolution considers the dimensions of the Brazilian territory and the transportation difficulties faced by some communities in certain regions of the country. In practice, it is not unusual see more than one public hearing held in different municipalities regarding the same project, which are good practices of companies and environmental agencies. It facilitates the dissemination of information and enlarges public consultation opportunities.

Article 3 of CONAMA Resolution No. 009/87 states that the public hearing will be conducted by a representative of the environmental agency who, after an objective exposition of the project and its EIS nontechnical summary, will open discussions with stakeholders present at the hearing. At the end of each public hearing, brief minutes will be prepared.[102] All written and signed documents, which are handed to the chair of the public hearing, will be attached to the minutes of the public hearing. Thus, written documents regarding the project and its EIA will be part of the environmental permitting procedure and should be taken into consideration in the decision-making.

Regarding the conduct of the hearing, Machado makes important points. Article 3 does not specify who will present the project and the EIS. However, the exposition of the project must be objective, which means impartial. Therefore, it would not be appropriate an exposition of the project made by the proponent of the project, which would certainly be partial and subjective.[103] In practice, the proponent of the project explains the project itself and the consultants who prepared the EIS present their findings, such as the projects impacts.

As far as partiality is concerned, the consultancy who prepares the EIS and presents it is partial since the consultancy is chosen and hired directly by the project proponent, which means that they have an interest in approving the EIA. In the same vein, the environmental agency suffers political pressure from high instances of the government, especially considering massive projects, which are usually very controversial. Therefore, the agency might not be as neutral as it should be. Furthermore, agencies conduct the public hearings, but they do not have the techniques to deal with this kind of meetings, which can be long, stressful and confrontational. It can also be large meetings. For instance, it is estimated that 6,000 people attended one of the public hearings of the Belo Monte Dam.[104] There are other situations, such as the arrogance of those who present in hearings, that can intimidate participants.[105]

Thus, it is necessary to have a neutral professional to facilitate the discussions and conduct dialogues among stakeholders, proponent of project, and environmental agency, in a productive manner. Having a neutral facilitator would improve the quality of the hearing and the credibility of stakeholders in what is being presented. This model is adopted by some jurisdictions. For instance, in Quebec, Canada, an independent office is in charge of organizing public consultations (the *Bureau the Audiences Publiques sur l'Environnement* or BAPE). BAPE is composed of commissioners appointed by the minister of the environment.[106] Once the public hearing is requested by citizens or an association and it is accepted by the minister of the environment, the president of BAPE nominates a consultation committee and a chairperson.[107] The role of the committee is to carry out the public hearing and everything related to it, such as meetings with project proponents and with those who requested the hearing, as well as mediation of discussions during the hearing; the committee receives written documents regarding the project and its EIS and prepares a report to the environmental minister, which is posted online for public consultation.[108] In Quebec, these public hearings might last three to five days, which demonstrates the importance of having enough time to discuss the EIS.[109]

Some scholars have evaluated the practice of public hearings in Brazil. One study analyzed forty public hearings, which happened from 1988 to 1996 in the most populous state of Brazil, São Paulo.[110] The study points out that although public hearings are important—because people had the opportunity to vent what they think, and it was easier, therefore, to reach a solution when conflicts were apparent—the way public hearings occurred was inadequate. The information provided was partial and insufficient time was allotted for effective discussion.[111]

Furthermore, means of presenting EISs and location of public hearings are problematic. Language used to present EISs is very technical.[112] Thus, laypersons do not fully comprehend what is explained. Therefore, they cannot provide their views on something that they do not understand. What is more, due to the potential of there being too much confrontation during the hearings, in general, there is not a consensus on the proposed project.[113] The geographical location chosen to hold a public hearing is another issue. Sometimes they happen in a city, kilometers away from the area that will be directly affected by the project. Therefore, the most affected people will not be able to participate.

Sanchez highlights the importance of holding prehearing meetings. He states that these meetings should be carried out by the proponent of the project.[114] In Brazil, some enterprises conduct meetings with communities prior to public hearings, establishing informational stands and hotlines and/or distributing pamphlets to communities explaining the proposed projects and its impacts. These are good practices; however, the project proponents still have bias. Consequently, information is not provided with neutrality. This was confirmed in the Volta Grande Project. During interviews in the state of Pará, several stakeholders, who participated in meetings regarding the installation of the Volta Grande Project, stated that the company was just advertising the project, instead of explaining it and its impacts.[115]

Although public hearings present some problems, they also serve an important role that cannot be disregarded. However, it seems that this is not the understanding of the Brazilian legislative branch. In 2015, a draft law was proposed by the Senate: PLS 654/2015. It proposes a special and brief environmental permitting procedure for infrastructure projects considered strategic and of national interest. According to the draft law, the environmental agency would have sixty days to analyze the EIS and sixty more days to issue a decision. The entire environmental permitting procedure would last around eight months.[116]

In this procedure public hearings would not happen; instead, a period of thirty days would be set to clarify questions that stakeholders might have. Senator Romero Jucá, author of draft law PLS 654/2015, stated that, due to lack of adequate conditions, environmental agencies spend around four years analyzing an EIS, which results in financial loss for private companies. Reflecting upon this statement, if environmental agencies do not have adequate conditions to issue a decision in four years, it seems unlikely that these agencies will do so in a shorter period. This draft law demonstrates an inversion of values of some Brazilian representatives. Exactly when public participation should be more active (infrastructure projects on national interest), the draft law proposes its decrease.[117] As of the writing of this book, the draft law was approved by the Special Commission of National Development and was awaiting a vote in Congress.

Overall, CONAMA Resolution No. 009/87, which regulates public hearing in Brazil, has only six articles and many deficiencies. It raises many questions about who will expose the project and the EIS, as well as the minimum deadline for realization of the public hearing, and it only refers to the nontechnical summary instead of the EIS and its nontechnical summary. Aside from regulation weaknesses, practice has demonstrated that the occurrence of public hearings in Brazilian EIA procedures is not efficient either.

Therefore, proposals to adopt some important measures could improve public hearings in Brazil: (1) public hearings should be obligatory following the long procedure, proposed in chapter 4 (screening), for certain projects; (2) a minimum period of time must be established to make the EIS and its nontechnical summary available so stakeholders can prepare for participation in hearings; (3) written guidelines should be provided on the availability of EIS, such as where they will be available and when they can be consulted, prior to public hearings;[118] (4) a professional and neutral facilitator should be responsible for conducting the public hearing; (5) an independent review committee should be in charge of making sure that the public hearing(s) occur(s) in places accessible to those most affected and language used is appropriate for laypersons to understand the projects being presented and its impacts.

CONCLUSION

Public participation in environmental decision-making is recognized in several international documents such as the Aarhus Convention and the Escazú Agreement. A participative EIA procedure is subject to three conditions: transparency, access to information, and grounded decisions. The first part of the chapter analyzed transparency and access to information in light of EIA regulation and practice.

Transparency in EIAs is limited by industrial and trade secrets. The Federal Constitution states that an EIS shall be made public, and all persons have the right to receive information of private interest or of collective or general interest from public agencies, which shall be provided within the period established by law, subject to liability, except for information whose secrecy is essential to the security of society and of the state. Therefore, government should weigh in on deciding which information could be omitted from public consultation.

Access to information in EIA procedures implies in the disclosure of documents, the provision of enough time for consultation of these documents, and the utilization of accessible language.[119] Furthermore, several procedural measures are necessary to guarantee access to information during the EIA. It is necessary to acknowledge to the public that the EIS and its nontechnical summary are ready for consultation through different and effective channels of communication.

In Brazil, Article 4 of Federal Law No. 10650/2003 states that the project proponent must publish the request, rejection, or approval of EISs and nontechnical summaries in official gazettes. This practice does not reach the population. While CONAMA Resolution No. 001/86 states that the RIMA will be available for public consultation, this practice violates the Federal Constitution, which states that the EIS shall be be made public. Therefore, both EIS and RIMA should be subject to public consultation. Additionally, the language used in EISs and its nontechnical summaries are often too technical, which imposes a barrier for access to information.

The written comments period and public hearings are the two methods of public consultation in Brazilian EIA procedure. CONAMA Resolution No. 001/86 has some shortcomings concerning these two consultation practices. For instance, it does not set a limit of time for written comments as it is established in other countries. As per public hearings, the literature has highlighted its limits, such as there not being any real dialogue due to too much confrontation, those who present and conduct the hearings are often biased, and the use of technical language makes comprehension and discussions very difficult.

A set of measures would overcome the aforementioned shortcomings: (1) using alternative and different means of communication to inform the availability of EIS and its nontechnical summary prior to written comments and public hearings, such as television and radio announcements, pamphlets, and outdoor posts, among others; (2) establishing an independent committee in charge of preparing and conducting scoping workshops and public hearings; (3) requiring the independent reviewing committee ensure that language is appropriate and serve as a channel of communication among communities, environmental agencies, and project proponents.

NOTES

1. Andréa Zhouri, "Mapping Environmental Inequalities in Brazil. Mining, Environmental Conflicts and Impasses of Mediation," (working paper no. 75, Desigualdades.net-Int. Res. Network on Interdependent Inequalities in Latin America, 2014), http://www.desigualdades.net/Resources/Working_Paper/WP-Zhouri-Online.pdf.

2. See Philippe Sands et al., *Principles of International Environmental Law*, 3rd ed. (Cambridge: Cambridge University Press, 2012), 624.

3. Ibid.

4. UN Conference on Environment and Development, Rio Declaration on Environment and Development, Principle 17, A/CONF.151/26/Rev.1 (Vol. I), Annex I (Aug. 12, 1992).

5. Paulo Affonso Leme Machado, *Direito Ambiental Brasileiro* (Brazilian Environmental Law), 21st ed. (São Paulo: Malheiroso Editores, 2013), 298, 299.

6. Ibid. See also Janice Miller, "Implementation of Principle 10 of the Rio Declaration of the United Nations Conference on Environment and Development" (paper, Caribbean Forum: Shaping a Sustainable Development Agenda to Address the Caribbean Reality in the Twenty First Century, March 5, 2014) (on file with author).

7. Ibid.

8. Sands et al., *Principles of International Environmental Law*, 625.

9. United Nations Economic Commission for Europe, Convention on Access to Information, Public Participation in Decision-Making and Access to Justice in Environmental Matters, June 25, 1998, 2161 U.N.T.S. 447.

10. Press Release, Economic Commission for the Latin American and the Caribbean, Fourteen Countries Sign New Generation Agreement at UN Headquarters on Access to Information, Public Participation and Justice in Environmental Matters (September 27, 2018) last accessed December 6, 2019, https://www.cepal.org/en/pressreleases/fourteen-countries-sign-new-generation-agreement-un-headquarters-access-information.

11. Marcos A. Orellana, "Democracia Ambiental y Desarrollo Sostenible: Hacia un Instrumento Regional sobre Derechos de Acesso" (Environmental Democracy and Sustainable Development: Towards a Regional Instrument on the Right to Access to Information) (2014), https://accessinitiative.org/sites/default/files/orellana_iaf2014.pdf.

12. Christopher Wood, *Environmental Impact Assessment: A Comparative Review*, 2nd ed. (London: Routledge, 2003).

13. Leonardo Avritzer, *Participatory Institutions in Democratic Brazil* (Washington, DC: Woodrow Wilson Center Press, 2009).

14. Three forms of participatory designs happened in Brazil with the 1988 Federal Constitution: bottom-up, power sharing, and public ratification. The bottom-up form consists on low involvement of the government and strong public participation and decision-making. Power sharing is a design where "bottom-up participation meets states actors" through mandatory negotiation and deliberation. Finally, public ratification consists of public deliberation to approve or reject a policy. Avritzer, *Participatory Institutions*.

15. Zhouri, "Mapping Environmental Inequalities in Brazil."

16. Ibid., 18.

17. Jonathan A. Fox and Libby Haight, "Mexico's Transparency Reforms: Theory and Practice," in *Government Secrecy* ed. Susan Maret (Bingley, UK: Emerald, 2011), 353.

18. Ibid.

19. Ibid.

20. Machado, *Direito Ambiental Brasileiro*, 298.

21. Directive 2011/92/EU on the Assessment of the Effects of Certain Public and Private Projects on the Environment, 2014 O.J. (L 124) 1.

22. Constituição Federal (C.F.)(Constitution), art. 225 (Braz.).

23. Ibid., art. 5, XXXIII.

24. Édis Milaré, *Direito do Ambiente: A Gestão Ambiental em Foco* (Environmental Law: Environmental Management in Focus), 7th ed. (São Paulo: Editora Revista dos Tribunais, 2011), 497.

25. Decreto No. 88.351, de 1 de Junho de 1983, Diário Oficial da União (D.O.U.) de 03.06.1983 (Braz.).

26. Machado, *Direito Ambiental Brasileiro*, 298.

27. Constituição Federal (C.F.) (Constitution), art. 5(XXIX) (Braz.).

28. Daniela Marcos Barone, "A Proteção Internacional do Segredo Industrial" (The International Protection of the Industrial Secret) (master's thesis, Faculdade de Direito da USP, 2009), http://www.teses.usp.br/teses/disponiveis/2/2135/tde-19112009-133733/pt-br.php.

29. Cheryl Ann Bishop, *Access to Information as a Human Right* (El Paso: LFP Scholarly Pub., 2012), 21.

30. "In the first opinion of this kind from an international human rights tribunal, the Inter-American Court on Human Rights [. . .] The case began in 1998 when the Chilean government refused to release information regarding a major U.S. logging project in Chile to a Chilean environmental organization." In reference to the case where the European Court of Human Rights ruled the decision recognizing access to information as a human right, "the case involved a request by the Hungarian Civil Liberties Union for access to a complain submitted to the Hungarian Constitutional Court." Bishop, *Access to Information*, 1–3.

31. American Convention on Human Rights, November 21, 1969, 1144 U.N.T.S. 143.

32. Cheryl Wasserman, "Enforcement of Environmental Impact Assessment Requirements," (paper, Ninth International Conference on Environmental Compliance and Enforcement, June 20, 2011), http://inece.org/conference/9/proceedings/57_Wasserman.pdf.

33. Ibid.

34. The World Bank, *Environmental and Social Framework* (Washington, DC: The World Bank, 2017): 99, § 19, http://pubdocs.worldbank.org/en/837721522762050108/Environmental-and-Social-Framework.pdf.

35. Nick Okello et al., "The Doing and Un-doing of Public Participation During Environmental Impact Assessments in Kenya," *Impact Assessment & Project Appraisal* 27, no. 3 (September 2009): 217, 221.

36. For SISNAMA see Chapter 2, EIA in Brazil, table 2.1.

37. Lei No. 10.650, de 16 de Abril de 2003, Diário Oficial da União (D.O.U.) de 17.11.2003 (Braz.).

38. Lei No. 6.938, de 31 de Agosto de 1981, Diário Oficial da União (D.O.U.) de 02.09.1981(Braz.).

39. Resolução CONAMA No. 006, de 24 de Janeiro de 1986, Diário Oficial da União (D.O.U.) de 17.02.1986 (Braz.).

40. Machado, *Direito Ambiental Brasileiro*, 297.

41. Antonio F.G. Beltrão, *Aspectos Jurídicos do Estudo de Impacto Ambiental (EIA)* (Legal Aspects of Environmental Impact Assessment [EIA]) (São Paulo: MP Editora, 2008), 46.

42. Ibid.

43. Wasserman, "Enforcement of Environmental Impact Assessment Requirements."

44. Machado, *Direito Ambiental Brasileiro*, 299.

45. Decreto No. 88.351, de 1 de Junho de 1983, Diário Oficial da União (D.O.U.) de 3.06.1983 (Braz.).

46. Machado, *Direito Ambiental Brasileiro*, 299, 300.
47. Ibid.
48. Ibid.
49. Constituição Federal (C.F.) (Constitution), art. 225, § 1, IV (Braz.).
50. Resolução CONAMA No. 001, de 23 de Janeiro de 1986, art. 8, Diário Oficial da União (D.O.U.) de 17.02.1986 (Braz.).
51. Lucivaldo Barros Vasconcelos, "O Estado (In) Transparente: Limites do Direito a Informação Socioambiental no Brasil," (The Lack of Transparency of the State: Limits of Social and Environmental Information in Brazil) (PhD dissertation, Universidade de Brasília, 2008), 276, http://repositorio.unb.br/handle/10482/4167?mode=full.
52. Resolução CONAMA No. 001/86, art. 11, §1.
53. Barros, "O Estado (In) Transparente," 274–279.
54. Ibid.
55. "Terras Indígenas no Brasil" (Indigenous Lands in Brazil), Instituto Socioambiental, last accessed December 11, 2016, https://terrasindigenas.org.br/pt-br/brasil.
56. "Module 3.2.: What is Public Involvement?," Environmental Impact Assessment Open Educational Resources, United Nations University, last accessed December 6, 2019, http://www.raymondsumouniversity.com/eia-local/page129.htm.
57. "Module 3.4 Principles of Public Involvement," Environmental Impact Assessment Open Educational Resources, United Nations University, last accessed December 6, 2019, http://www.raymondsumouniversity.com/eia-local/page131.htm.
58. See Miriam Cuppen et al., "Public Participation in EIA and Attitude Formation," *Impact Assessment & Project Appraisal* 30, no. 2 (April 2010): 63–74, http://dx.doi.org/10.1080/14615517.2012.660348.
59. See Luiz Enrique Sanchez, "Avaliação de impactos para licenciamento ambiental no Brasil: Modernização fundamentada no aprendizado." (Environmental Assessment for Environmental Licensing in Brazil: Updates Based on Lessons Learned), (PowerPoint presentation, Environmental Licensing: Reality and Perspectives, Brasília, November 5, 2015), http://www.mpf.mp.br/atuacao-tematica/ccr4/dados-da-atuacao/eventos/Seminarios/Ppt0000055.pdf.
60. Elisabeth Conrad et al., "Hearing but Not Listening? A Participatory Assessment of Public Participation in Planning," *Environment & Planning, Government & Policy* 29, no. 5 (October 2011): 761, 773.
61. Shaun Fluker, "The Right to Public Participation in Resources and Environmental Decision-Making in Alberta," *Alberta Law Review* 52, no. 3 (2015): 567, 579.
62. Meinhard Doelle and A. John Sinclair, "Time for a New Approach to Public Participation in EA: Promoting Cooperation and Consensus for Sustainability," *Environmental Impact Assessment Review* 26, no. 2 (March 2006): 185–187.
63. Council on Environmental Quality (CEQ) Regulations for Implementing the Procedural Provisions of the National Environmental Policy Act, 40 C.F.R. § 1503.1 (b) (2015).

64. "Module 3.2.: What is Public Involvement?."

65. 40 C.F.R. § 1503.3 (2015).

66. Ibid., § 1503.4.

67. Code of Practice, Consultation in Ontario's Environmental Assessment Process, Ontario (website), last updated January 2014, https://www.ontario.ca/document/consultation-ontarios-environmental-assessment-process.

68. IBAMA, *Avaliação de Impacto Ambiental: Caminhos para o Fortalecimento do Licenciamento Ambiental* (Environmental Impact Assessment: Pathways to Enhance the Environmental Licensing) (2016),last accessed December 6, 2019, http://www.ibama.gov.br/phocadownload/noticias/noticias2016/resumo_executivo.pdf.

69. Machado, *Direito Ambiental Brasileiro*, 302, 303; Decreto No. 40, October 30, 2012, art. 90, Diario Oficial (D.O.) (Chile).

70. Machado, *Direito Ambiental Brasileiro*, 302, 303.

71. Instrução Normativa do IBAMA No. 184, de 17 de Julho de 2008, Diário Oficial da União (D.O.U.) de 18.07.2008 (Braz.); Instrução Normativa do ICMBIO No. 14, de 27 de Outubro de 2011, Diário Oficial da União (D.O.U.) de 28.10.2011 (Braz.).

72. 40 C.F.R. pts. 1500-1508 (2015).

73. Ministério Público da União 4a Câmara de Coordenação e Revisão—Meio Ambiente e Patrimônio Cultural), *Deficiências em Estudo de Impacto Ambiental* (Shortcomings of Environmental Impact Assessments) (2004).

74. Ibid.

75. 40 C.F.R. § 1.502.9 (2015).

76. CARE Food and Nutrition Security, Pathways Program: The Talking Book (last visited), https://worldbankgroup.sharepoint.com/sites/gsg/CDD/Lists/Calendar/Attachments/209/CARE%20Talking%20Book%20Innovation%20Brief.pdf.

77. Conrad et al., "Hearing but Not Listening?," 773.

78. Ebenezer Appah-Sampong, "Public Hearing Within the Environmental Impact Assessment Review Process," in *UNEP Environmental Impact Assessment Training Resource Manual*, 2nd ed. (UNEP, 2002), http://www.unep.ch/etu/publications/16%29%2085%20to%2091.pdf.

79. Milaré, *Direito do Ambiente*, 501.

80. Appah-Sampong, "Public Hearing Within the Environmental Impact Assessment Review Process," 87.

81. Ibid.

82. CEQ Regulations for Implementing the Procedural Provisions of the National Environmental Policy, S. 1506.6.

83. Machado, *Direito Ambiental Brasileiro*.

84. Department of Electricity Development and Ministry of Water Resources, *Manual for Conducting Public Hearings in the Environmental Impact Assessment Process for Hydropower Projects* (Kathmandu, Nepal: Private Sector Hydropower Development Project, 2004), http://www.doed.gov.np/documents/Manual-for-Conducting-Public-Hearings.pdf; John Glasson et al., *Introduction to Environmental Impact Assessment* 3rd ed. (London: Routledge, 2012) 150, 151.

85. Department of Electricity Development and Ministry of Water Resources, *Manual for Conducting Public Hearings.*

86. Glasson et al., *Introduction to Environmental Impact Assessment*, 150, 151; Joe Weston, "EIA, Decision-making Theory and Screening and Scoping in UK Practice," *Journal of Environmental Planning & Management* 43, no. 2 (March 2000): 185, 186.

87. W. E. Westman, *Ecology, Impact Assessment and Environmental Planning* (New York: Wiley, 1985).

88. Department of Electricity Development and Ministry of Water Resources, *Manual for Conducting Public Hearings.*

89. James L. Creighton et al., *Involving Citizens in Community Decision Making: A Guidebook* (Washington, DC: Program for Community Problem Solving, 1994).

90. Markus Gunther et al., "Many Roads May Lead to Rome: Selected Features of Quality Control within Environmental Assessment Systems in the US, NL, CA and UK," *Environmental Impact Assessment Review* 62 (January 2017): 250; Appah-Sampong, "Public Hearing Within the Environmental Impact Assessment Review Process," 89.

91. Angus Morrison-Saunders and Gerard Early, "What Is Necessary to Ensure Natural Justice in Environmental Impact Assessment Decision-making?," *Impact Assessment & Project Appraisal* 26, no. 1 (2008): 29.

92. CONAMA Resolution entered into force in July 28, 1990.

93. Interviews with anonymous citizens and public officials involved with the environmental permitting of the Volta Grande Project, in Altamira, Senador José Porfírio, and Belém, Pará State, Brazil, July 2–9, 2015.

94. Milaré, *Direito do Ambiente.*

95. Machado, *Direito Ambiental Brasileiro*, 304.

96. Antonio Herman de Vasconcelos e Benjamin, "Os Principios do Estudo de Impacto Ambiental como Limites da Discricionariedade Administrativa" (The Principles of the Environmental Impact Assessment as Limits of the Administrative Discretion), *Review Forense* 317 (1992): 25.

97. Machado, *Direito Ambiental Brasileiro*, 304.

98. Milaré, *Direito do Ambiente*, 305; Machado, *Direito Ambiental Brasileiro*, 304.

99. Machado, *Direito Ambiental Brasileiro*, 305.

100. Ibid., 306.

101. Resolução CONAMA No. 009, de 03 de Dezembro de 1987, Diário Oficial da União (D.O.U.) de 05.07.1990 (Braz.).

102. Resolução CONAMA No. 001/86, art. 4.

103. Machado, *Direito Ambiental Brasileiro*, 305, 306.

104. Ministério Público Federal, "Altamira: Audiência Pública sobre Belo Monte Marcada por Protestos" (Altamira: Protests Happened During Public Hearing to Present Belo Monte), International Rivers, September 15, 2009, 1:09 p.m., https://www.internationalrivers.org/node/1759.

105. Ibid.

106. Luis Enrique Sanchez, *Avaliação de Impacto Ambiental: Conceitos e Métodos* (Environmental Impact Assessment Concepts and Methodologies), 2nd ed. (São Paulo: Oficina de Textos 2013), chap. 16, E-BOOK, https://ebooks.ofitexto.com.br/product/avaliao-de-impacto-ambiental-2-ed

107. Ibid.

108. Ibid.

109. Ibid.

110. Ibid.

111. Ibid.

112. Barros, "O Estado (In)Transparente," 276.

113. Sanchez, *Avaliação de Impacto Ambiental: Conceitos e Métodos*, chap. 16.

114. Ibid.

115. Interviews with anonymous citizens and public officials involved with the environmental permitting of the Volta Grande Project, in Altamira, Senador José Porfírio and Belém, Pará State, Brazil, July 2–9, 2015.

116. Rodrigo Baptista, "Avança Projeto que Acelera Liberação de Licenças Ambientais para 'Obras Estrategicas'" (Proposal of Granting Faster Environmental Licenses Authorizing Implementation of Governmental Strategic Projects Moves Forward), Agência Senado, November 26, 2015, 10:26 a.m., http://www12.senado.leg.br/noticias/materias/2015/11/25/avanca-projeto-que-acelera-liberacao-de-licencas-ambientais-para-obras-estrategicas.

117. See "Licenciamento Ambiental mais Rápido para Obras de Infraestrutura será Votado no Plenário" (Faster Environmental Licensing will be Subject to Vote at the Senate), Agência Senado, November 26, 2015, 10:26 a.m., http://www12.senado.leg.br/noticias/videos/2015/11/licenciamento-ambiental-mais-rapido-para-obras-de-infraestrutura-sera-votado-em-plenario.

118. Machado, *Direito Ambiental Brasileiro.*

119. The World Bank, *Operational Manual*, Operational Policy 4.01 (January 1999, revised April 2013), https://policies.worldbank.org/sites/ppf3/PPFDocuments/090224b0822f7384.pdf.

Chapter 8

Decision

The EIA requires decisions in each of its stages. Although this view is not always clear in practice and EIA stages are so interrelated that decisions are not apparent or formal, in fact from screening to the authorization of project, passing through scoping, identification of mitigation measure, and finally authorization, decisions are made. During the screening, decision is taken whether an EIA is necessary; in scoping, it is decided which impacts will be assessed; and during review, a decision is taken whether the EIA is approved or rejected.[1]

This chapter discusses the last stage of the EIA procedure: the decision to approve the EIA and, consequently, the authorization of the project. Approval of EIAs is different from approval of the project. In systems where EIAs happen within a permitting procedure, such as in the Brazilian system, the approval of the EIA signifies that it followed its guidelines and the provisions of the pertinent legislation as well as the quality of assessments, methodologies, and mitigation measures accounting for its approval.

Due to the procedural nature of EIA legislation, decision-makers can authorize projects even though they are proven to be very harmful to people and the environment.[2] Furthermore, they are not obligated to follow mitigation measures indicated in EISs.[3] The IAIA has emphasized that one major shortcoming of EIA procedures is the fact that it is poorly linked to decision-making.[4] Furthermore, one of IAIA's recommendations to all countries during the World Summit on Sustainable Development in Johannesburg, in 2002, was to strengthen impact assessment integration into decisions.[5]

Polonen points out that the procedural nature of an EIA is not a problem when EIAs are strongly connected to decision-making and integrated to environmental permitting procedure, when authorizing agencies must incorporate substantive requirements (to protect the environment) as conditions of the permit.[6] In fact, the obligation to incorporate environmental protective measures into permits is the major benefit of integrated systems. However, there will still be a great level of discretion to decision-makers since they can still approve projects independently of the severity of anticipated impacts.

As the EIA is a procedure that involves technical knowledge, analysis of alternatives, and consideration of public input, there is normally tension in decision-making among scientific, political, and public opinion. This is because the EIA is not an instrument which exists to protect the environment per se; instead it serves to acknowledge the potential impacts of projects in order to provide informed decisions.[7] The EIA is a planning instrument which assesses economic activities and its impacts on natural resources and people's lives. Therefore, it aims to better accommodate the different equation of destruction versus production.[8] In other words, ultimately, a cost-benefit analysis is made on implementing a project by finding the best social and environmental mitigation measures or concluding that the social and environmental impacts are unbearable.[9] Therefore, EIA "comprises three parts: politics, public and science" and "a successful EIA requires that all three be in balance."[10]

Derani states that the decision at the end of the EIA procedure is a "legally oriented political decision"[11] since the outcome is that a political choice is made to approve or deny a project. Science has a secondary role in assisting political decisions oriented by regulations and legislation.[12] Although Derani is right in relation to practice, ideally neither science nor the affected community's input should have a secondary role. These factors should have equal weight with political reasons.

There is a trade-off in approving/denying projects.[13] Thus, conflicts arise when it is difficult to conciliate different priorities (communities, developers, and governments).[14] Mitigation and compensation measures partially assist in resolving such conflicts; however, they will not return the natural environment to the status quo. It is difficult to strike the balance between environmental, social protection and economic growth. However, this is possible by adopting a procedure that is truly democratic.

In developing countries, economic interests in many situations supersede social and environmental concerns. Therefore, three questions arise: (1) How should economic forces and social and environmental concerns be weighed in order to have equilibrated decisions? (2) Who should decide on the approval of projects at the end of the EIA procedure? (3) How to balance scientific knowledge in the EIA decision-making and subsequent authorization/rejection of projects?

SCIENCE AS OPPOSED TO LEGITIMATE DECISIONS

EIA involves science because it is necessary to collect baseline environmental data prior and probably also during the installation and operation of a proposed project to understand potential impacts; for example, modeling air emissions or classification of impacts in different levels of occurrence (e.g., high, medium, or low probability). Including public participation in a procedure that contains highly technical information is challenging. The US EPA states that

> Good environmental decision-making is a complex process which requires both a careful review and assessment of relevant science and a thoughtful application of social values. Science alone does not bring many solutions. It does not predict with 100% certainty the impacts as well as costs to mitigate them. Thus, even the most science-based decision must be supported by social values.[15]

One common problem is that information provided by experts receives more weight from EIAreviewers and authorizing agencies than public input and experience. Yet, indigenous populations, fishermen, and long-term residents can contribute valuable knowledge. Gregory and McDaniel state that this source of alternative knowledge, which is different from Western scientific information, can inform the decision to authorize projects.[16] For example, in ecosystem restoration projects severely impacted by dams, local residents can provide useful information of ecological processes prior to disruption. Or residents and traditional communities can raise concerns on species that were not considered in the EIA.[17] Therefore, both the scientific community and traditional knowledge are important and should complement each other.

For instance, in deciding the best alternative for lowland irrigated agriculture, in the Phitsanulok irrigation project in Thailand, a multi-criteria technique was used—combining sustainability factors, economic considerations, and farmers' preferences.[18] Geographical information system was used to evaluate sustainability criteria and a two-stage field survey incorporated peoples' views in the decision-making.

EIAs aid decision-making, but they only predict impacts. Sometimes there are high levels of uncertainty. In this regard Gregory and McDaniel state that decision-makers should differentiate categories of uncertainty; for instance, the severity of impacts, capacity of the environment to absorb the impacts, whether mitigation measures are appropriate, and capacity of proponent to mitigate the impacts. All these should be well-documented and balanced during decision-making, which will analyze how each of these and other topics will be managed during the implementation of the project.[19]

Uncertainties should definitely be documented. However, when the level of uncertainty is high and the probability of impacts could be disastrous, projects should not go forward. In light of the Precautionary Principle, "Where there are threats of serious or irreversible damage, lack of full scientific certainty shall not be used as a reason for postponing cost-effective measures to prevent environmental degradation."[20] Therefore, in cases where EIAs reveal that impacts are uncertain with probability of severe impacts, projects should not be approved. Alternatively, other designs, locations, or technologies capable of decreasing the level of uncertainty of impacts could be proposed.

THE WEIGHT OF ECONOMIC FORCES AND POLITICAL WILL AS OPPOSED TO ENVIRONMENTALLY SOUND DECISIONS

In developing countries, where poverty is apparent and systemic, regulations are weak; when norms have strong social and environmental provisions they are not enforced, it is even more challenging to find the right balance among science, economic needs and local community consensus. What role could the law play in avoiding, at least partially, the economic interests that override serious environmental and social welfare of local communities? Literature on decision-making processes recognizes this as an area where further research is needed.[21]

Particularly relevant are collaborative decisions, which are very complex and difficult to achieve. Gregory and McDaniel state that "The underlying behavioral questions—how well the deliberative process is understood, how thoughtfully the input is provided, and how meaningfully outputs are considered (in terms of process and outcome linkages)—remain areas of frustration to theorists and practitioners alike."[22] The Global Development Research Center identifies the following criteria for "better" environmental decision-making:[23]

- Legitimacy—decisions that were constructed in collaboration with stakeholders, they are recognized as fair, trusted, and accountable. It results from inclusive and empowering processes.
- Equity—different interest was identified and included in the final decision on equal footing. It also addresses social justice and ecological rights and how they will be managed.
- Efficiency—development should provide more benefits than costs. In other words, "sustainability gains outweigh all the costs of implementation." This cost-benefit analysis needs to integrate stakeholders' views.
- Effectiveness—environmental decisions should take into account the natural environment and equilibrate with social needs.

Transparency should be added to the above list of characteristics. Decisions should be well-grounded and justified, clear and transparent, available to all, who want to access and, if necessary, challenge it in court. The possibility of questioning a decision is also a principle of democracy and a pillar of the access to information and public participation stated in Principle 10 of the Rio Declaration.[24]

While there is no defined standard (at least, not yet) or law capable of preventing political will from overriding serious social and environmental impacts—since it involves complex decisions which should make the balance of growing economically and minimizing impacts—certain substantive criteria would assist in reaching more socially and environmentally sound decisions. For instance, the adoption of most environmentally friendly alternatives as were required by Netherlands until 2010 was best practice.[25] It was mandatory to develop an environmentally friendly alternative to projects. In other words, the EIA aimed to find the most environmentally friendly technology, sites, timing, and processes.[26] However, this provision was removed from Netherland's EIA legislation.

Polonen states that, within the EU EIA Directive, the fact that it is not mandatory that decisions include EIA findings is a major weakness. Decision-makers should only take into consideration EIA findings and the mitigation measures proposed.[27] He proposes that the EU EIA Directive be strengthened with the inclusion of more substantive provisions. For example, the directive could have the following language: "No authority is empowered to grant a permit for the implementation of a project if the assessment indicates that the project would have a significant adverse impact on the environment."[28] This would be the main rule. However exceptional circumstances would rely on the following provision: "Permit can be granted if the Member State considers that the project must, in absence of alternative solutions, be carried out for imperative reasons of overriding public interest."[29] Although this is a good attempt to limit decisions that are not environmentally sound, these provisions would not limit such decisions because having this provision on exceptions would open space for governments to include all projects of its interest in the exceptional provision.

Decisions should be based on human rights. A substantive criterion capable of orienting decisions on approval of projects would be an assessment by decision-makers as to whether the project would disrespect the human rights of those who will be affected—specifically, their rights to housing, access to water, labor, and others. Further studies would be necessary to delimitate these criteria. Although provisions alone do not guarantee respect to people's rights, a regulation stating that authorization of projects cannot violate human rights would demonstrate a good faith effort to recognize and protect these rights.

Furthermore, the capacity of project proponents would be an important component to orient decisions. Paragraph 22 of the 2012 IFC Policy on Environmental and Social Sustainability states that the "IFC will only finance investment activities that are expected to meet the requirements of the Performance Standards within a reasonable period of time."[30] The capacity of project proponents is an important fact to consider when deciding on its implementation. Especially considering that some large-scale activities underestimate costs of implementation during authorization process.[31] In other words, the costs are higher than predicted and there is great risk that project proponents will not implement all the mitigation and compensation measures initially proposed. Therefore, an assessment of the capacity of implementation is an important criterion to decisions.

Analysis of a company's social and environmental performance could also be a substantive criterion to justify authorization/rejection of projects. In this regard the United Nations Guiding Principles on Business and Human Rights states that, in the context of conflict-affected areas, states should "[deny] access to public support and services for a business enterprise that is involved with gross human rights abuses and refuses to cooperate in addressing the situation."[32] However, in developing countries this is unlikely to be implemented, especially in relation to large-scale projects. Existence of substantive criteria explicitly stated in law or regulation on the approval of projects is important because it decreases the level of discretion of authorizing agencies/governments while additionally it gives stakeholders more leverage to challenge such decisions in court.

Inclusion of local affected communities' views in the decision-making of massive and potentially destructive projects as real dialogue is also capable of limiting governments' decision on the authorization of such projects. Furthermore, in democratic countries environmental decisions should be legitimate and not imposed on communities. Decisions such as relocation of communities to install a project on their land might drastically affect their way of living, their ties with the land, and interactions with each other. Therefore, in such cases affected communities' participation should be part of a process that leads to consensus or quasi-consensus. In this context Goodland states that, "While not perfect, FPIC is a vast improvement on using force in development or imposing involuntary conditions on impacted people."[33] Practical recommendations are provided within the next sections.

WHO DECIDES ON THE APPROVAL/ REJECTION OF PROJECTS SUBJECT TO EIA?

The literature defines the traditional approach of decision-making processes as top-down, DAD ("Decide, Announce, Defend"), or DEAD ("Decide,

Educate, Announce and Defend").[34] Such approaches are no longer accepted in democratic countries.[35] However, in practice they still exist. Modern theories propose participatory decision-making, known as MUM ("Meet, Understand, Modify") or POP (Public Owns Project").[36] MUM and POP are more appropriate and consistent with the role of EIA in aiding decisions over the implementation of projects which significantly affects people's lives and the environment.[37]

Furthermore, participation is only effective if it is capable of influencing the final decision-making in some way: by approving/rejecting the implementation of a project or changing its design and/or technology; choosing another alternative to development, including the "no project" alternative; the selection of another site for its implementation or agreeing on mitigation measures capable of minimizing the social and environmental adverse impacts.[38] Therefore, collaborative decisions have been gaining considerable attention in the literature and are considered best practice.[39] In this regard, the International Association for Public Participation developed a spectrum of public participation, which is transcribed in figure 8.1.[40]

Figure 8.1 demonstrates several levels of public engagement. EIA decision-making normally encompasses only the consultation level. However, collaborative decisions seem to be more appropriate in democratic countries, because it better accommodates different positions and wills concerning the implementation of potentially harmful projects. Therefore, it leads to accountable and transparent decisions.[41] Conversely, the "empower" level of public engagement, meaning "we will implement what you decide," might be too much power in the hands of the people, ultimately creating conflicts difficult to resolve among local affected people and state or national interests.

It is good practice to separate approval of the EIA on its quality, as determined by experts, from political decisions made by elected officials, such as the Minister of the Environment.[42] Joseph identifies four different groups who issue decisions based upon an EIA: (1) independent committee appointed by government, (2) elected officials based on independent committee, (3) stakeholders in consensus or near consensus (FPIC), and (4) elected officials based on stakeholders' input. All methods have strengths and shortcomings. For instance, independent committees might not be directly accountable to the public. However, the independence is important to guarantee neutral decisions.[43]

The process of decision-making varies among jurisdictions. In the United States, the agency responsible for the project (public projects) or for issuing the authorization or permit (private projects) is responsible for carrying out and approving the EIA. The fact that the agency approves its own EIA is highly criticized by practitioners and the literature.[44] In the UK, the authority

Increasing Impact on the Decision →

	Inform	Consult	Involve	Collaborate	Empower
Public Participation Goal	To provide the public with balanced and objective information to assist them in understanding the problem, alternatives, opportunities and/or solutions.	To obtain public feedback on analysis, alternatives and/or decisions.	To work directly with the public throughout the process to ensure that the public concerns and aspirations are consistently understood and considered.	To partner with the public in each aspect of the decision including the development of alternatives and the identification of the preferred solution.	To place final decision making in the hands of the public.
Promise to the Public	We will keep you informed	We will keep you informed, listen to and acknowledge concerns and aspirations, and provide feedback on how public input influenced the decision.	We will work with you to ensure that your concerns and aspirations are directly reflected in the alternatives developed and provide feedback on how public input influenced the decision.	We will look to you for advice and innovation in formulating solutions and incorporate your advice and recommendations into the decisions to the maximum extent possible.	We will implement what you decide.

Figure 8.1 IAP2 Spectrum of Public Participation. *Source*: International Association for Public Participation, https://cdn.ymaws.com/www.iap2.org/resource/resmgr/pillars/Spectrum_8.5x11_Print.pdf.

responsible for land planning is responsible for decisions on the approval/rejection of projects. In other jurisdictions, a council of ministers issues the decision.[45] In any case, the final decision is made by the government, either by an individual (a minister) or by a collegiate (several ministers or other elected officials).[46] Although a minister or elected official has the final say, he/she must do so keeping the balance of science, environment and economic goals. All these aspects must be acknowledged consistent with public concerns and input made during the procedure.

In order to guarantee accountable and equitable decisions, it is important that they are unbiased and made in conjunction with those who will be mostly affected.[47] Some mechanisms are important to guarantee neutral and substantial decisions.[48] The social control, through public participation, judicial and administrative controls, and independent committees, aims to ensure that decisions are impartial, comply with legislation and policies, and consider the public's views.

THE ROLE OF INDEPENDENT REVIEWING COMMITTEES/OFFICES

The Netherlands' Commission for Environmental Assessment (NCEA)

In the Netherlands, there are two different EIA procedures: the simplified and the full-fledged EIA. The latter is submitted for revision to the NCEA—a permanent committee in charge of reviewing complex, governmental, as well as all projects subject to the Dutch Nature Conservation Act.[49]

The NCEA is composed of experts who are independent from the government. Therefore, its technical decisions on the quality of the EISs are not influenced by political or administrative considerations.[50] Furthermore, the NCEA is not responsible for issuing the authorization or permit to implement the project, which also ensures its neutrality.[51] For each EIS, the NCEA sets up a working group composed of a chairperson, a secretariat and experts, who might be hired specifically for the EIS in question.[52]

By law, the NCEA must issue an advisory report on its review of the EIS. However, in the initial stages of an EIA, such as screening and scoping, the NCEA might act voluntarily or by request from the competent authority. Additionally, after the NCEA's revision, whether it is necessary to supplement the EIS or not, the NCEA has no obligation to further review the EIS, considering the information supplemented by its developer.[53]

The role of the NCEA is important due to its independence from political decision-making. However, it is not free from weaknesses. The fact that the NCEA is only obligated to issue a report when the EIS is completed is a weakness since the NCEA does not participate in the screening decision and most importantly in the inclusion of themes in the TOR or guidelines. Thus, the NCEA reviews only a completed document (the EIS), which is almost the end of the procedure. Moreover, in order to prepare its reviewing report, the NCEA is not required to consider the public's comments; however, it can do so, if requested by the competent authority.[54]

Public input should naturally be included in the NCEA report since communities can provide important information on the location where the project is planned to be implemented. In a comparative study on EIA systems in the UK and the Netherlands, Arts concludes that "in the Netherlands participation is seen as being of less importance for influencing EIA, but important for its overall acceptance."[55] This assertion might explain why public participation is not primarily included in the NCEA report.

The Canadian Review Panel

In Canada, the EIA is conducted by either the Impact Assessment Agency of Canada or the review panel. Impact assessment of projects, regulated by the Nuclear Safety and Control Act or the Canadian Energy Regulator Act, is referred by the Minister of Environment to an integrated review panel.[56]

Relevant to this analysis is the impact assessment conducted by a review panel when the minister is the opinion that the referral is of public interest.[57] The minister would consider the following: the extent of adverse effects caused by the project within federal jurisdiction or its direct or incidental effects, public concern related to the effects of the project in question, opportunities for cooperation with any jurisdiction that is related to the assessment of the environmental effects of the proposed project, any project's adverse impacts on the indigenous peoples of Canada.[58]

The agency appoints as member of the review panel one or more persons who are unbiased and free from any conflict of interest relative to the proposed project, who have knowledge or experience relevant to the project's anticipated effects, or have knowledge of the interests and concerns of the Indigenous peoples of Canada that are relevant to the assessment.[59] Documents produced by the review panel must be available for public consultation and hearings are held before the panel concludes its final report. The entire procedure must take up to 600 days.[60] The minister will consider the review panel report as well as public inputs during the procedure. Under Section 51 of the Impact Assessment Act

> A review panel must, in accordance with its terms of reference, (a) conduct an environmental assessment of the designated project; (b) ensure that the information that it uses when conducting the environmental assessment is made available to the public; (c) hold hearings in a manner that offers the public an opportunity to participate meaningfully, in the manner that the review panel considers appropriate and within the time period that it specifies, in the impact assessment; (d) prepare a report with respect to the impact assessment . . . submit the report with respect to the impact assessment to the Minister; and (f) on the Minister's request, clarify any of the conclusions and recommendations set out in its report with respect to the impact assessment.[61]

The World Bank Inspection Panel

In the 1990s, the World Bank Group adopted Social and Environmental Safeguards to ensure that the implementation of projects it finances would comply with the goals of sustainable development.[62] Accordingly, even developing countries, which do not have strong environmental regulations and laws, would need to comply with the World Bank Social and Environmental Safeguards and ensure that the project respects social and environmental guaranties.[63]

The Inspection Panel (IP) is a post-facto mechanism and therefore it is different from the previous reviewing committees; however, its function and impartiality are important for the discussion of ensuring social control over decisions made by the WB. The IP is an independent grievance mechanism for people and organizations who believe projects financed by two of the World Bank Group branches—the International Bank for Reconstruction and Development (IBRD) and International Development Association (IDA)—do not comply with its own policies and procedures and therefore would adversely affect their lives.[64] The panel is composed of three members appointed by the board of executive directors who serve a nonrenewable five-year period.[65] Some requirements guarantee the independence of the panel: their members cannot work at the World Bank Group at the end of the period they served as members of the panel and World Bank staff members may only be appointed to the IP two years after the termination of their contract with the World Bank. The IP has a permanent secretariat, independent from the Bank's staff and management.[66]

Once the IP receives a complaint, it reviews whether the management team, responsible for implementation of the project, complied with WB policies and procedures. According to the IP Operational Procedure, it has the following accountability functions:

> A. It provides a forum for people, including those who are often poor and vulnerable, to seek recourse for harm which they believe result from Bank-supported operations. As such, the Panel is a "bottom-up" or citizen-driven accountability mechanism that responds to grievances and demands for redress. This promotes more inclusive and sustainable development by giving project-affected people a greater voice in Bank-financed projects that impact them. B. It provides an independent and impartial assessment of claims about harm and related non-compliance with Bank policies as a check-and-balance for the Board and other concerned stakeholders. This contributes towards institutional learning and helps to improve development effectiveness of World Bank operations.[67]

When a complaint is filed, the IP verifies the eligibility criteria and carries out a preliminary assessment by analyzing the request for inspection as well as the response of the WB staff management, responsible for the project object of the complaint.[68] Then, it recommends to the board of executive directors whether an investigation should be carried out or not. If the board decides that an investigation should occur, the IP will investigate and present its findings to the board, which will consider the necessary measures to be taken by the Bank. The final report as well as other documents are available on the IP's website.[69]

The IP is an accountability mechanism that inspired other development institutions to create similar mechanisms, such as the International Finance Corporation Compliance Advisor Ombudsman Office (CAO), the Asian Development Bank's Accountability Mechanism, and the Inter-American Development Bank's Independent Consultation and Investigation Mechanism, among others.[70]

WB was the first international institution (outside of the European Union) to create a citizen-driven mechanism to verify and monitor the implementation of its own procedure and policies.[71] Hunter states that by creating the IP, the Bank allows a "citizen a direct role in monitoring the institution's performance against normative standards. It gives affected communities direct access to the highest political levels of the Bank."[72] However, the process has some flaws. Bridgeman and Hunter note that one weakness of the IP procedure is that the WB's management staff, who could be found in noncompliance with the Bank's policies and procedures, are also responsible for implementing the IP's recommendations.[73] Furthermore, Sarkar states that the findings of the IP are merely administrative rather than a binding decision.[74] According to Sarkar, by creating a general lending practices protocol between the WB and its members, noncompliance would be considered a breach of international law, which should be enforceable at the international level.[75]

Although the IP's process has some flaws, it is an important mechanism. The public availability of the IP's findings, as well as management staff reports and executive boards' decisions, ensure transparency and pressure the management staff to follow the IP's recommendations. Otherwise, its nonobservance might impact the WB's credibility.[76]

The Office of the Compliance Advisor Ombudsman (CAO)

The CAO is an independent accountability mechanism of the private arms of the World Bank Group: the International Finance Corporation (IFC) and the Multilateral Investment Guarantee Agency (MIGA).[77] As the IP, the CAO is a post-facto mechanism. The CAO reports directly to the president of WB and is mandated to "Address complaints from people affected by IFC/MIGA projects (or projects in which those organizations play a role) in a manner that is fair, objective, and equitable; and enhance the environmental and social outcomes of IFC/MIGA projects (or projects in which those organizations play a role)."[78]

To carry out its mandate, the CAO has three functions: ombudsman or dispute resolution, compliance, and advisory. In the same vein as the IP, to guarantee its independence and impartiality, the chairperson of the CAO—the vice president—must never serve the World Bank Group in any position, even after leaving the CAO.[79] The vice president will serve in this role for a period of three to five years, renewable by mutual consent. Furthermore, CAO staff can only work in the IFC/MIGA after a period of two years from the termination of their contract with the CAO.[80]

Any individual, group, or organization who believes they are affected or will be potentially affected by a project funded by IFC/MIGA can file a complaint. Once a complaint is received, the CAO screens it according to its eligibility criteria.[81] The positive screening decision is posted on its website, while decisions on the ineligibility are communicated to the complainant in written format.[82]

Once a complaint is received and deemed eligible, complainants can choose whether they want to resolve the issue under the dispute resolution function or have the compliance appraisal process straight away.[83] In any case, if the complainant decides for the dispute resolution function and the case is not resolved, it is automatically transferred to compliance function, which will assess whether an investigation is merited.[84]

The CAO Operational Guidelines states that "as a non-judicial, non-adversarial, neutral forum, CAO's approach provides a process through which parties may find mutually satisfactory solutions. This role facilitates an approach that ensures equitable treatment of participants in a dispute resolution process."[85] The CAO uses several approaches to settle the dispute, which ranges from informal consultations to formal mediations.[86]

If CAO decides that a case merits an investigation, it will hire specialists with appropriate expertise to form a compliance panel. When hiring experts, the CAO will take into consideration their knowledge, independence, and impartiality.[87] During the investigation process, the compliance panel will conduct a technical review of the investment to verify whether IFC complied with its own sustainability framework. This comprises review of EIAs, environmental management systems, and environmental and social action plans, among others. The findings of the compliance investigation are sent to the management team of IFC/MIGA for their factual review and comments. Finally, the report is sent to the president of the Bank who will assess both the response of the management team and the findings of the CAO.[88] In both functions, dispute resolution and compliance, the CAO monitors either the agreement resulted from the settlement of the dispute or the compliance investigation report.[89] This is a strength of the CAO procedure since it not only identifies the noncompliance but also monitors it until IFC/MIGA turns back into compliance.[90]

All is done in a transparent approach since the monitoring documents are available for public consultation on its website. However, monitoring is a flaw of the CAO process, especially in the compliance role. Complainants normally do not have a chance to provide input in the compliance report nor to participate in the development of an action plan to remediate the situation.

On the contrary, in the Project Complaint Mechanism of the European Bank for Reconstruction and Development, complainants do have a chance to comment on the action plan proposed by the Bank.[91]

In sum, the CAO is a good model of independent accountability mechanism. Its role in dispute resolution is to resolve social and environmental impacts of projects financed by IFC/MIGA in an independent and neutral fashion. The CAO procedures do contain flaws, such as lack of participation of complainants in the design of an action plan and the fact that its report is not binding, therefore, increasing its chances to be vulnerable and politicized.[92] Despite these weaknesses, having an independent office responsible for resolving disputes and investigating noncompliance with social and environmental policies is a great step toward implementation of sustainable projects in developing countries. Therefore, it is a model worth following.

BRAZILIAN EIA DECISION-MAKING

Who Approves EIA and Issues Permits Authorizing Projects in Brazil

In Brazil, the environmental agency (federal, state, or local) in charge of conducting the environmental permitting procedure approves the EIA, and it might authorize the project or not. If the project is authorized, the agency issues the preconstruction permit.[93] In some states, such as São Paulo, Bahia, and Minas Gerais, the decision to issue environmental permits is made by committees comprised of members of civil society, public defenders, and professional associations, among others.[94] Their decision is based on the environmental agency's final report on the EIS.[95] The existence of these committees is a good initiative to legitimize the decisions.[96] However, such committees are not composed of experts and they might not assess the EIS in great detail. Furthermore, they typically follow the recommendations in the report from the environmental agency on the approval of projects.[97]

Beltrão states that, in the Brazilian system, environmental matters are excessively concentrated within environmental agencies. In his view, not only environmental agencies but all public administration are also bound by environmental legislation.[98] According to Beltrão, a system similar to the United States—where agencies from different sectors (water, energy, oil) are responsible for conducting the procedure and issuing the environmental permit—would be more appropriate.[99] Environmental agencies would then revise norms and standards to be observed and also monitor their implementation.[100]

Sanchez also criticizes the fact that in Brazil the final decision on approval/rejection of projects is made by environmental agencies.[101] Contrary to the system in Brazil, the Regional Centre for Economic Development, Transport and the Environment in Finland is responsible for conducting and reviewing EIAs but does not issue the permit.[102] Similarly, in Canada and the Netherlands, agencies responsible for reviewing the EIA do not issue permits.

In fact, environmental agencies suffer political pressure from frequent instances of government efforts to implement strategic and massive projects in Brazil. For example, IBAMA head Abelardo Bayma Azevedo resigned in January 2011 alleging he was being pressured to issue the environmental permit for the Hydroelectric Belo Monte power plant. His predecessor, Roberto Messias Franco, also left the position due to divergent opinions regarding the dam.[103]

Furthermore, environmental agencies normally do not have enough qualified staff to deal with a great number of EIAs. For instance, in 2014, IBAMA had 428 environmental analysts to review all EISs filed at the agency.[104] Data demonstrate that, from 2003 to 2014, the number of procedures conducted by IBAMA increased from approximately 480 a year to 1,800 annually.[105] In this regard, Thomaz Miazaki de Toledo, who was the director of permitting procedure at IBAMA in 2015, has stated that it is necessary to strengthen the organizational structure of environmental agencies.[106]

Obviously, there are drawbacks to concentrating environmental permitting procedures (and EIA reviews) with the environmental agencies. However, these agencies have specialized personnel and, theoretically, they should be impartial when analyzing the approval of projects since they are not affiliated to any specific sector (water, energy, health, etc.). The Brazilian system of concentrating environmental permitting procedures with environmental agencies could be efficient for noncontroversial or noncomplex projects, of which there are less or no political pressure and public consultation would be necessary but not as robust as in massive projects, also slated to be implemented in sensitive locations. For these projects, special attention should be given to educate communities to participate and reach consensus. An independent and totally neutral body of experts would assist in this task.

As for impartiality, Brazilian environmental agencies' staff are normally public servants; although, in the state of Pará, SEMAS has a number of consultants. The auditing court, however, has criticized the use of any outsourced reviewers of EIA, stating that "Environmental Impact Assessment happens within the Environmental Permitting Procedure; therefore, it should not be conducted by external consultants."[107] In other words, the review of EIAs should only be carried out by public servants.

Public servants are not subject to termination of their work contracts, unless there are serious reasons prescribed by law. Moreover, they are subject to legal proceedings regarding whether their decisions cause significant harm to any person.[108] However, in practice, decisions on the authorization of major projects remain questionable.[109] In this sense, a report of the Brazilian auditing court concluded that decisions on EIA procedures excessively focus on issuing permits rather than considering the real impacts of projects or assessing the efficiency of mitigation measures (follow-up monitoring measures).[110]

The decision to approve the Belo Monte Dam is a perfect example. Although there were several legal actions against the issuance of the permit—as well as a recommendation from the IACHR stating that the project implementation should be suspended due to a lack of FPIC of indigenous communities—the permit was still issued.[111] The problem in granting permits is not just a legitimizing one but also public credibility regarding decisions made in the approval of major projects.[112]

Machado states that, considering that the entire procedure aims to provide a basis for decision-makers, "it is important to achieve an independent structure for environmental agencies at all levels."[113] Moreover, reviewing agencies must have as much independence as possible. For instance, the same level of independence that judges have.[114]

Regarding the substance of decisions CONAMA Resolution No. 001/86 does not shed light on how decisions should be made. For instance, which aspects should be considered in an EIA decision. Furthermore, there is not any provision on the obligation to make decisions available for public consultation. Therefore, it is necessary to enhance CONAMA Resolution No. 001/86 and include guidelines on decisions as well as procedural obligation to make them available in print in rural areas or places difficult to access and on the internet.

Reasons for Approving Potentially Harmful Projects

Practice has demonstrated that decisions on major projects are made with the argument of bringing benefits to local population in detriment to stakeholders' consensus or quasi-consensus.[115] On the other hand, the voice of the people through protests and rallies has been pushing the government to consider affected communities' social and environmental rights. For instance, plans for implementation of the Belo Monte Dam started over thirty years ago; the project was subsequently modified and redesigned due to social pressure. While the original project would flood 1,225 square kilometers of land, including indigenous lands, the already implemented dam's reservoir is two-third smaller and no indigenous land is flooded. Although implementation of the current dam has many flaws, the change of the design over the years proves that social pressure was important and had a role in this change.[116]

The Brazilian government normally justifies the installation of projects ignoring many rejecting voices (civil society, researchers, affected communities) based on the "fighting poverty" argument. In this same way, the private sector often plays a role that does not belong to them. For example, in the case of Volta Grande Project, Belo Sun affirmed that they were bringing improvement to the community in the Volta Grande since they were distributing food baskets to the residents.[117] What is more, according to Belo Sun, communities will be impacted if they do not implement the project because it would stop development in the region.[118]

The discourse of the State Environmental Secretary José Alberto da Silva Colares, during the Public Hearing of the Volta Grande Project, was also that the conditions in which people live in the Volta Grande were precarious and the goal was to improve their lives.[119] Additionally, one person responsible for analyzing the EIS of Volta Grande said, "I wrote a favor opinion in relation to the issuance of the permit of this project because of the conditions of the people who live there"[120]—meaning the development would ameliorate their lives. However, as noted in the public hearings documented in the permitting procedure, there were objections to the project and serious concerns of the *Ministério Público* regarding its cumulative impacts in the Volta Grande area, its impacts on indigenous and other traditional communities living in the region.

In this vein, a public defender of the state of Pará, who knows the Volta Grande Project well, stated that "In environmental permitting procedures in Brazil there is no real dialog and the justifications are always the same: people are poor, they have nothing, consequently, any improvement brought by projects is beneficial to them."[121] However, it is necessary to respect their culture, the way they live, their environment, their right to be close to the river, close to each other, and their will. Obviously, there are regions of Brazil that need attention and need to grow economically. Yet it is not the implementation of projects at any cost and without proper communities' consultation and consensus that will fill the gap of many years of state absence.

Another prosecutor from the state of Pará said that "[The Volta Grande Project] is a very interesting case because in the case of implementation of Belo Monte, the state can justify that the dam will generate energy to the country. Thus, there could have public interest in its installation. However, in the case of the Volta Grande Project it is mere economic power that justifies its implementation because gold mining is not in the public interest."[122]

Aside from political forces, some practical and legal aspects contribute to this situation. There is weak social control;[123] CONAMA Resolution No. 001/86 does not provide any guidance on the transparency of decision and parameters to ground agencies' decisions. In this regard, Part 1505 of the US CEQ Regulations is dedicated to decision-making. Section 1505.2 states that once a decision is made, the agency shall

> 1 State the decision; 2 Identify alternatives considered by the agency in reaching its decision, specifying the alternative or alternatives considered environmentally preferable. An agency may discuss preferences among alternatives based on relevant factors including economic and technical considerations and agency statutory missions. An agency shall identify and discuss all such factors, including any essential considerations of national policy, that the agency balanced in making its decision and state how those considerations entered into its decision; 3 State whether the agency has adopted all practicable means to avoid or minimize environmental harm from the alternative selected, and if not, why the agency did not. The agency shall adopt and summarize, where applicable, a monitoring and enforcement program for any enforceable mitigation requirements or commitments.[124]

As mentioned in chapter 4, the Federal Constitution states that principles of lawfulness, impartiality, morality, publicity, and efficiency guide governmental decisions.[125] Moreover, decisions shall be reasoned.[126] Therefore, these principles orient environmental agencies on decision-making regarding approval or rejection of projects. However, even with the constitutional and legal obligations, some decisions on projects remain obscure.

In relation to disclosure of decisions, in the United States, the CEQ Regulations Section 1505 states that decisions should be published in the record of decisions.[127] Moreover, Section 1506.1 states that no action can be taken until the decision is published.[128] In Brazil, a good practice was established in IBAMA Executive Order No. 184/2008.[129] It created the Environmental Permitting Procedure Online System (in Portuguese *Sistema Informatizado de Licenciamento Ambiental* [SISLIC]). The SISLIC aims to provide more transparency to environmental permitting procedures. Documents such as EIS, public hearing transcriptions, EIA technical decisions, and monitoring reports are available online for download and public consultation.[130] Although this is not the reality of many state environmental agencies, it seems a good step toward the implementation of transparent measures by disclosing important EIA documents.

In sum, transparency is crucial to guarantee social control over decisions on approval or rejection of projects. Federal EIAs are approved by the environmental agency, which is responsible for issuing the permit and consequently authorizing the project. This system has strengths and shortcomings. Authorization of projects that potentially affect delicate environments and communities should not be decided by the environmental agency. Conversely a process that leads to consensus and therefore FPIC would be more appropriate. To better contextualize this proposal, it is necessary to analyze FPIC in Brazil and who has FPIC rights to finally conclude that it is legally possible to extend FPIC to other communities, aside from indigenous.

FREE, PRIOR, AND INFORMED CONSENT (FPIC) IN BRAZIL

As defined in chapter 1, FPIC is the right of indigenous and tribal peoples to be informed prior to a decision of authorization of projects or policies that might affect their land, in order to provide or withhold their freely and informed consent. ILO Convention No. 169 contains explicit articles on FPIC. Although Brazil was one of the latest countries in the Americas to ratify the convention,[131] Decree No. 143 of June 20, 2002, implemented it domestically.[132]

ILO Convention No. 169 applies to indigenous, tribal, and semi-tribal peoples. Article 1 states the following:

> 1. This Convention applies to: (a) tribal peoples in independent countries whose social, cultural and economic conditions distinguish them from other sections of the national community, and whose status is regulated wholly or partially by their own customs or traditions or by special laws or regulations; (b) peoples in independent countries who are regarded as indigenous on account of their descent from the populations which inhabited the country, or a geographical region to which the country belongs, at the time of conquest or colonization or the establishment of present state boundaries and who, irrespective of their legal status, retain some or all of their own social, economic, cultural and political institutions. 2. Self-identification as indigenous or tribal shall be regarded as a fundamental criterion for determining the groups to which the provisions of this Convention apply.[133]

The convention applies to indigenous peoples and quilombolas, who are recognized ethnic minorities in Brazil.[134] Yet more broad interpretations exist in relation to traditional communities. In relation to indigenous peoples, Article 231 of the Federal Constitution states that

> Indigenous peoples shall have their social organization, customs, languages, creeds and traditions recognized, as well as their original rights to the lands they traditionally occupy, it being incumbent upon the Union to demarcate them, protect and ensure respect for all their property . . . Paragraph 3—Hydro resources, including energetic potentials, may only be exploited, and mineral riches in Indigenous land may only be prospected and mined with the authorization of the National Congress, after hearing the communities involved, and the participation in the results of such mining shall be ensured to them, as set forth by law.[135]

Therefore, the Federal Constitution recognizes that activities in indigenous lands will only be authorized with the authorization of the National Congress and after indigenous communities are consulted. Moreover, the Federal Constitution recognizes the rights of quilombolas to the land they traditionally occupy. Article 68 states that "To the remaining of quilombo communities who are occupying their lands, their definitive property is recognized. The State shall issue the competent land title."[136] They are understood as being "tribal peoples whose social, cultural and economic conditions distinguish them from other sections of the national community, and whose status is regulated wholly or partially by their own customs or traditions or by special laws or regulations," as defined in Article 1 of ILO Convention No. 169. Thus, they have been recognized as peoples who have the right to FPIC.[137] The Brazilian court has defined quilombolas as tribal peoples. Therefore, their rights to their lands must be assured.[138] The court, in a case which recognizes property rights of quilombolas according to ILO Convention No. 169, issued the following decision:

> The State cannot neglect the constitutional protection as one of the fundamental objectives of the Federative Republic of Brazil, which is to promote the well-being of all, without prejudice of origin, race, sex, identity and any other form of discrimination (CF/88, art. 3o, IV) including, therefore, communities reminiscent of quilombos. . . . I am confirming [the] understanding that to establish policies to tackle discrimination to traditional lifestyles of indigenous and tribal peoples, when the enactment of Legislative Decree No. 143/2002, which ratifies the ILO Convention No. 169, states in Article 14 that the rights of property and ownership of the people [dealt in this Convention] regarding the land they traditionally occupy.[139]

This is not only the Brazilian but also the Colombian and Ecuadorian interpretation on the application of ILO Convention No. 169, which also have legislation recognizing afro-descendants the right to FPIC.[140] Furthermore, the IACHR condemned Suriname for not carrying out consultation prior to authorization of a hydroelectric power plant in the land where the Saramaka people live, recognizing that

> A.1 [T]he Saramaka people are one of the six distinct Maroon groups in Suriname whose ancestors were African slaves forcibly taken to Suriname during the European colonization in the 17th century . . . the Court considers that the members of the Saramaka people make up a tribal community whose social, cultural and economic characteristics are different from other sections of the national community, particularly because of their special relationship with their ancestral territories, and because they regulate themselves, at least partially, by their own norms, customs, and/or traditions. . . . Hence, this Tribunal declares that the members of the Saramaka people are to be

> considered a tribal community, and that the Court's jurisprudence regarding indigenous peoples' right to property is also applicable to tribal peoples because both share distinct social, cultural, and economic characteristics, including a special relationship with their ancestral territories, that require special measures under international human rights law in order to guarantee their physical and cultural survival.[141]

Overall, in Brazil and its neighboring countries, there is no doubt that activities and projects developed on indigenous and quilombola lands are subject to FPIC as a condition to their authorization. Despite these requirements, it is difficult to enforce this right. Instituto Socioambiental states that FPIC in Brazil, in some cases, works as an instrument to stop projects, in others it is an efficient negotiation space and in other cases it is not respected at all.[142]

In light of above two points deserve attention: (1) whether other traditional (or nontraditional) communities (apart from quilombolas and indigenous) have the right to FPIC in relation to the implementation of projects in the land they occupy and (2) how to interpret the requirement of consent for this purpose.

Do Local Affected Communities Have (or Should Have) FPIC Rights?

Brazil's large geographic area includes African, native indigenous, and European influences; the number of traditional communities and peoples in Brazil is still unknown.[143] Consequently, it has diverse cultural traditions. The Project New Social Cartography in the Amazon (in Portuguese, *Nova Cartorgrafia Social da Amazonia*) developed maps and extensive research identifying several traditional communities in Brazil.[144] The research resulted in twenty reports, each volume dealing with one traditional community, including Faxinais in Paraná, Fundos de Pasto in Bahia, artisanal fishermen in South Bahia, river-dwellers and artisans in Itaquera, Gaspar, Barreira Branca, and São Pedro—Jauaperi River, Roraima, and Amazonas.[145] Many traditional communities live not only in the Amazon but also throughout the country. Because these communities often live in distant and/or rural areas, they are normally affected by projects as these areas are not occupied by large populations providing space for installation of major companies or are attractive to development due to the availability of natural resources.

Decree No. 6040/07 sets the National Policy of Sustainable Development of Traditional Communities and Peoples. Article 3 defines traditional communities and peoples as "groups culturally different, who recognize themselves as such, have their own social organization, occupy and use territories and natural resources as condition for their cultural, social, religious, economic and ancestral reproduction, utilizing knowledge, innovation and practices learned and inherited by tradition."[146]

Duprat states that the main criteria of Decree 6040/07 to identify traditional communities and peoples was inspired by ILO Convention No. 169.[147] Therefore, traditional peoples and communities are those who "retain some or all of their own social, economic, cultural and political institutions."[148] In her view, all traditional communities and peoples are considered in ILO Convention No. 169. Therefore, they have the right to provide or withhold their consent.[149]

A study written by Instituto Socioambiental and Rainforest Foundation Norway corroborates the interpretation that ILO Convention No. 169 should apply to all traditional communities and peoples, based on the fact that Decree 6040/07

> describes three elements of tribal and traditional communities and peoples stated in Article 1 of ILO Convention No. 169, which are the existence of social, cultural and economic conditions different from other sectors of national society; the presence of social organization which is directed totally or partially by their own rules and traditions and the self-identification, which means that the social group has its tribal identity.[150]

Although there is no Brazilian legislation recognizing FPIC rights to traditional communities and peoples, recognition is slowly gaining new heights in the development debate.

In response to this growing recognition, the Federal *Ministério Público* filed a legal action against the state of Amazonas regarding construction of a port.[151] In this case, the judge determined that all acts related to the construction of a port in Amazonas were canceled until riverbank communities were heard, according to Articles 6 and 15 of ILO Convention No. 169. The judge stated that

> Traditional Communities are not defined in a closed list. It is any human group with peculiar characteristics, management systems which respect nature cycles, utilization of knowledge inherited by past generations, who identify themselves as such. The law shall not reject their identity on the grounds that these communities and peoples shall fill out forms defined by others who do not belong to these groups.[152]

Furthermore, FPIC should be expanded to all local affected communities, traditional or not. Self-identification as tribal or indigenous groups is the criterion utilized to determine traditional communities (indigenous, quilombolas, and others), as it is stated in Article 1 Paragraph 2 of ILO Convention No. 169.[153] However, other communities might not identify themselves as traditional. For example, those who live in shantytowns in Brazil. In these cases, criteria to determine FPIC rights could be scale of projects, cumulative impacts, communities' dependency on natural resources, fragile ecosystems, among others.

Critiques to recognizing FPIC rights to all local affected communities (especially nontraditional ones, since in this case there is no international convention defining their rights) could be that the right to development would be undermined whether all local affected communities and peoples would have a more powerful voice in the participation process. As mentioned in chapter 1, fundamental rights have equal value.[154] In this sense, the right of local communities to participate in a process to reach their consensus about decisions that affect their lives or their activities of fishing, hunting, socializing with each other would not be less important than the right of a state to develop and grow economically. Both, rights of local affected communities and right to development, are stated in the Federal Constitution. The right to development is also stated as one of the objectives of the Federal Republic of Brazil.[155]

Fiorillo points out that economic activities shall be oriented by social justice, respecting the principle of protection of the environment as it states in Article 170, VI, of the Federal Constitution:

> The economic order, founded on the appreciation of the value of human work and on free enterprise, is intended to ensure everyone a life with dignity, in accordance with the dictates of social justice, with due regard for the following principles: VI—environmental protection, which may include differentiated treatment in accordance with the environmental impact of goods and services and of their respective production and delivery processes.[156]

All local affected communities should have FPIC rights when the implementation of large- or medium-scale projects, with significant cumulative impacts, would severely impact local communities' lives and/or require their resettlement. Other criteria could be applied when projects require FPIC, such as those slated to be implemented in fragile ecosystems.

Although FPIC does not mean veto power, it does require meaningful dialogue with the objective of reaching consensus or quasi-consensus. Goodlands states that FPIC does not need to reach majority consensus.[157] FPIC does not mean that communities in relation to projects will always be prevalent. What must be ensured, however, is that the process involves meaningful participation of affected communities at early stages of EIA procedure.

Distinction between Consultation, Consent, and Consensus

Recalling previous discussions on consultation, consent, and consensus,[158] consent is not an isolated event but a collaborative process to achieve consensus.[159] There are different positions in relation to consent, such as those who state that consent would be pointless when communities who have FPIC rights do not hold a veto power in relation to policies or projects that will affect their lives and those who maintain that consent is part of the consultation process and, therefore, a mere formality which would not interfere in the decision-making.[160]

Duprat states that both arguments are equivocated.[161] First, because ILO Convention No. 169 requires consent in relation to projects that affect indigenous, tribal peoples and other traditional communities.[162] Thus, FPIC cannot serve as mere consultation—meaning that indigenous and traditional communities would be heard, however their views would not necessarily be taken into account. Second, because in a society, limited groups of people cannot have exclusive powers in decisions that involve interests beyond their own.[163]

As Grabner recognizes, ILO Convention No. 169 does not provide veto power to indigenous and tribal peoples.[164] However, it does impose that a consensual solution should be reached, especially considering that the decision of the state cannot be imposed on them.[165] According to Instituto Socioambiental and Rainforest Foundation Norway FPIC is "mainly a space for discussion and influence for developing coordinated decisions in a context that excludes the possibility of imposition in any of the parts."[166] The ILO has corroborated this assertion, stating that FPIC does not mean veto powers.[167] Admittedly, this view has merit because in environmental decisions there are many values to be considered and many distinct situations, therefore, it is necessary to have collaborative decisions and not unilateral ones.

In the Saramaka case, the IACHR distinguished between consent and consultation when it stated that "consent applies when big projects would potentially result in loss of territory or serious impacts."[168] According to Duprat, consent is absolute in these situations, and it would lead to vetoing of projects. For all other situations, consent should be relative, yet binding. Objections should be incorporated into the decision-making and the project should be totally or partially modified.[169] While loss of territory is an objective criteria, the occurrence of serious impacts is relative and difficult to define in practice.

FPIC is an important political instrument, more powerful than mere consultation. Additionally, in certain circumstances consultation is not enough, such as when traditions, social groups, and way of living are threatened. Ideally FPIC to all local affected communities should aim to reach a consensus or quasi-consensus. A collaborative decision should be sought and pursued and not imposed on communities.

According to the US EPA, "Studies of public participation have emphasized the importance of developing a climate of cooperation among participants so that they are willing to consider alternative values and viewpoints and the possibility of surrendering a portion of their individual autonomy for the collective good. Trained facilitators can often be effective in helping to create such a climate."[170] Hence, reaching consensus or quasi-consensus is more feasible when participation is meaningful.

FPIC in South America

It is under countries' independent regulations of ILO Convention No. 169 to determine whether a FPIC procedure is deemed necessary.[171] In South America, some countries outline procedures in their legislation such as Colombia and Venezuela. Bolivia's legislation prescribes that procedures and deadlines to obtain FPIC will be defined on a case by case basis.[172]

To make communities' participation meaningful it is necessary that these communities have assistance in explaining to them the impacts of projects. Independence of technical support is of crucial importance because it will guarantee credibility of the consultation procedure.[173] The law of indigenous peoples in Venezuela and the regulation of Bolivia states that technical support should be provided to communities who will participate in the consultation process.[174,175]

As per the area of application of FPIC processes, regulations in South America normally define that activities within traditional communities' lands require consultation processes such as in Ecuador and Bolivia.[176] Colombia's legislation states that when indigenous or black communities live within the area of influence of the project or activity, the project proponent will include these communities in the EIA and will consutl them.[177]

The Volta Grande Project

To facilitate understanding the FPIC process proposed in this book, consider the Volta Grande Project. Riverbank residents, fisherman, and other traditional and perhaps nontraditional (e.g., artisanal miners) communities live in the area of the project.[178] In an interview in Altamira, a member of a nongovernmental organization mentioned that "In the absence of the State, people who live in Ressaca Village organize themselves to help each other. In the past, when someone got sick, villagers got together and collected money to take that person to Altamira, to be treated in a hospital"[179] (there is no hospital in Ressaca village and the trip to Altamira by boat is costly and long—it takes around three hours). This statement demonstrates that they organize themselves with a sense of solidarity to resolve their problems because the state has left them behind (there is no potable water, sewage, health center, or police station in Ressaca village).

As mentioned in many interviews, communities in the villages were divided in relation to the installation of the project. Some even claimed that many participants in the public hearing were children who signed the list of presence to give the impression that there were many attendees. Due to different opinions the sense of solidarity among people at least in Ressaca village had disappeared.

The Volta Grande case could have had a different outcome, although one not finished yet, if a well-conducted participatory process had been in place in the early stages of the EIA. Communication was weak; inhabitants of Ressaca village affirmed that, even after two public hearings, they did not know if they would be relocated or not.[180] In relation to disclosure of the EIS, legal intervention was necessary to guarantee local communities' access to the document, and the language was too technical to be clearly understood by laypersons. These flaws could have been avoided by having an early understanding of the project and its impacts. Considering peoples' views and concerns earlier (during scoping) and constructing a true dialogue along the procedure could have allowed a consensus or quasi-consensus to be reached in relation to its approval or rejection. Instead the environmental permit was issued and later suspended due to lack of consultation to indigenous peoples who live in the area of the project.

In summary, it is clear that indigenous communities and quilombolas have FPIC rights in relation to projects that potentially affect their lands and lives. This understanding has been applied to other traditional communities in Brazil. The argument that whether nontraditional communities had FPIC would supersede the right to development of the country does not stand because the right to development has equal value than the right to life and the right to an ecologically balanced environment. Therefore, affected nontraditional communities should have FPIC rights.

CRITERIA FOR EIA REVIEW AND DECISION-MAKING

The procedural nature of EIAs concerns practitioners regarding effectiveness since authorizing agencies are not obligated to reject a project based on EIA findings or to integrate proposed mitigation measures into decision-making. However, the definition of effectiveness is relative. Bond states that there is not a common understanding of effectiveness.[181] From a public participation point of view an effective EIA would be a procedure that truly included stakeholder's participation in all or most of its steps. Furthermore, Destra states that impact assessments are effective when they are participative and "when views and perceptions of as many actors as possible are accounted for."[182]

Furthermore, an effective EIA is a tool that informs decision-makers on the positive and negative impacts of projects. This is achieved when EIA is conducted and reviewed by impartial parties, when it is transparent, and quality of data is accurate. It is necessary to have clearly defined criteria to review EIAs and therefore guarantee informed decisions.

Bond states that effectiveness is measured against control mechanisms. He adapted six control mechanisms identified by Ortolano, which influence EIA effectiveness:[183]

- Procedural control - administrative units set EIA requirements;
- Judicial control - courts are empowered to investigate and judge procedural non-compliance;
- Evaluative control - reviewing agencies assess EIAs and make recommendations to authorities who are responsible for authorizing projects;
- Development aid agency control - financing institutions require EIAs prior to disbursing the resources to support a given project;
- Professional control - EIA developers observe professional standards and codes of conduct;
- Direct public and agency control - communities and governmental agencies exercise control by pressuring to influence an EIA outside the aforementioned control mechanisms.

In Brazil, procedural and judicial control exist in relation to environmental permitting procedure, which encompasses the EIA. In relation to evaluative control, the agency that reviews the EIA also authorizes the project. Regarding aid agency control the National Development Bank (BNDES) is a major player in financing large-scale projects in Brazil. BNDES has its own social and environmental policies, aside from requesting borrowers comply with Brazilian laws.[184] Professional control is under private sector best practices. Finally, regarding direct public and agency control, as it was thoroughly explained there are some opportunities for intervening agencies and public participation in EIA procedures; however, participation in most cases is late and not meaningful. Based on the effectiveness control mentioned above Bond proposes the following EIA effectiveness control dimensions:[185]

1. Procedural compliance—did the EIA comply with the rules and regulations?
2. Completeness of EIA documents—are significant impacts covered in adequate detail?
3. Methods to assess impacts—are the methods used for prediction and evaluation appropriate?

4. Influence on project decisions—has the EIA had any influence on decisions, for example, design modifications, location changes, rejection of application?
5. Development of knowledge and learning—does the approach to EIA develop knowledge and enhance learning among stakeholders?
6. Comprehensive involvement—does the approach to EIA engage with members of the public and other stakeholders in a meaningful way at a point in time early enough to influence the conduct of the EIA?

Creighton states that "It is not enough just to reach a decision. If it is going to 'count,' a decision must be seen as legitimate. What makes a decision 'legitimate' is not only the substance of the decision but also the perception that the process by which the decision was made was fair, open, and democratic."[186] Moreover, "[. . .] what gives legitimacy to a decision made using public participation is the fact that the public is able to influence the entire process. Not every public participation program will result in a consensus building. But every decision will be made in a visible manner, with opportunities throughout the process for groups to influence the outcome."[187]

To ensure effective EIA review, the independent committee should ensure that public participation was conducted properly from the early stages of EIA; compliance with procedural requirements and substantive criteria would also decrease occurrences of political will being predominant on the decision on the approval of projects. Therefore, it would be necessary to amend Brazilian EIA regulation to include the following reviewing and decision-making criteria of projects requiring FPIC of local affected communities. Independent committees would ensure that

- Public participation was meaningful;
- Stakeholders were answered appropriately;
- Procedure complied with legal steps;
- Human rights were respected (check whether there is no human rights violation);
- Socially and environmentally friendly criteria were adopted;
- Local and social aspects were considered: Do people want the project? How do they see it? How do they want it? What are the best project's alternatives available? How will the project address social impacts? Are these measures enough? Does the company/project proponent have capacity (financial and technological) to implement these measures?
- Consensus or quasi-consensus was reached.

This would be the minimum criteria applied by both the independent committee and the authorizing agency. The formal opinion of the committee should address the above points in written format. Additionally, the decision authorizing the project should follow the same pattern. Therefore, the decision on the authorization/rejection of projects could be challenged in court based on the aforementioned criteria.

While FPIC does not mean veto power, in situations where consensus cannot be reached, a decision would be made by the authorizing agency responsible for issuing the permit. To protect communities and guarantee that decisions will not favor only economic forces, the law or regulation should include substantive criteria on decision-making such as those indicated in this chapter for reviewing EIAs.

PROPOSAL: EIA INDEPENDENT COMMITTEES IN BRAZIL

Independent committees, not only as impartial reviewers of procedures but also as mediators, would be a feasible solution in Brazil.[188] Having independent committees would remove the responsibility of issuing the permit from those in charge of reviewing EIAs. Thus, technical decisions would not mix with political ones and analysis of EIAs would not be concentrated in the environmental agency (or at least the most controversial or sensitive ones).

The committee proposed here would have the characteristics of existing independent committee, panel, and accountability mechanisms highlighted earlier in this chapter: independence, impartiality, and expertise. Regarding structure, the independent committee would have two main roles (similar to the CAO): stakeholders' engagement, which would coordinate communication with stakeholders by electing a facilitator who would conduct workshops and public hearings and EIA review.

The independent committee would have its fixed staff with expertise in facilitation and environmental science; however, they could hire experts to assist in specific projects according to the subject in question. Additionally, preference should be given to experts from the area slated to be affected by the project. Moreover, a regulation on the composition of the panel would assist to guarantee their independence.[189] Arend states that "a mechanism to ensure the quality and independence of the review is necessary to guarantee scientifically sound and reliable study."[190] Thus, an independent review committee would play this role.

In relation to legislative changes necessary to implement the measures proposed, a federal law should create the independent committee, which would have financial autonomy and would be placed under the Ministry of Environment and should be included in the SISNAMA structure. However, this committee would not be responsible for authorizing projects or issuing permits.[191]

Federal Law No. 140/11 delimitates the competence to conduct environmental permitting procedures. As EIAs happen within the environmental permitting procedure, Federal Law No. 140/11 would need to be amended in order to establish the competence for the full public participation procedure proposed in this book. Therefore, only one entity of the federation would be responsible for conducting the environmental permitting procedure and issuing the environmental license. However, the independent committee would be in charge of reviewing the EIA, providing its technical opinion, approving it and sending it to the appreciation of the environmental agency and consequent approval or rejection of the project.

Currently, Federal Law No. 140/11 is based on localization of projects. It should remain based on this criterion, except for certain projects, such as large-scale or medium-sized projects located in sensitive ecosystems such as the Amazon, projects requiring displacement and/or when cumulative impacts with other projects are highly significant. In these cases, the project would require a full public participation EIA in which the environmental permitting would be conducted by IBAMA and the EIA would be reviewed by the EIA independent reviewing committee. To illustrate this point, the Volta Grande Project environmental permitting would be conducted by IBAMA and its EIA would be reviewed by the EIA independent committee. This book does not provide a detailed list of projects that should be subject to the full EIA public participation; it merely provides some examples. Further research would assist in identifying projects' impacts and localization which would require in-depth public participation.

As per the experience learned in the case study presented here, it is clear that communities in the Amazon need special attention whenever projects would affect their lives. Therefore, a full public participation EIA is necessary in these areas and it would definitely benefit project proponents, local communities, and government. Legislation amendments, especially the competence and screening criteria, should be well-defined and subject to public consultation throughout the country.

In general, CONAMA Resolution No. 001/86 would need to adapt the changes for the full public participation EIA. For instance, it would need to include in its Article 3 that IBAMA (in general) or the EIA independent committee (under certain circumstances) would be responsible for approving EIAs. All the public participation measures (feasibility studies, workshops, public hearings, written comments) should also be incorporated into the same CONAMA Resolution. Currently, a different resolution regulates public hearings, CONAMA Resolution No. 009/87; however, it would be more practical that one resolution would deal with the EIA requirements including actions to engage the public.[192]

CONCLUSION

The EIA procedure aims to provide a basis for sustainable decision-making in development projects. It is considered best practice to separate the technical expert's decision on the quality of EIAs from governmental decisions on the necessity of implementing projects.

In some jurisdictions such as Canada and the Netherlands, independent review panels, who are not responsible for issuing permits, conduct and review EIAs and issue a report containing experts' evaluation of EIAs. Quasi-judicial grievance mechanisms implemented by international institutions such as the World Bank IP and the International Finance Corporation CAO also play an important role in ensuring that projects funded by these institutions comply with their own social and environmental policy procedures. The IP and CAO offer great examples for reviewing EIAs maintaining independence from political or private interests. However, they are post-facto mechanisms.

Furthermore, the necessity of having a team or a facilitator capable of conducting conversations among stakeholders is an important function of the CAO. This fact demonstrates that regarding sensitive issues such as impacts of projects, it is worthwhile to try as much as possible to resolve the issues rather than have a constant battle.

Although EIA is not a decision-making process per se, it is a tool to aid decisions. Thus, the question of who should decide on the approval of projects is relevant since often public participation is not taken into consideration and decisions disregard people's will even when traditions and cultures are threatened by decisions. FPIC should be extended to all local affected communities. It is a matter of justice and democracy to account for their consent in relation to decisions that impact their lives. In relation to traditional communities, the main criterion which defines "traditional" in ILO Convention No. 169, which is in summary, is that they organize themselves differently from society as a whole, embraces all traditional communities, aside from indigenous and quilombolas in Brazil.

The procedure proposed here does not aim to undermine in any manner FPIC right already guaranteed to indigenous and quilombolas (in Brazil). It just aims to expand this right. Moreover, guidance is proposed in order to facilitate the consultation process, which can be adapted to traditional communities' own manner of participation and decision-making. Therefore, it is not a closed and rigid procedure with inflexible rules and measures.

In Brazil the Federal Constitution and the Administrative Procedure Act provide general guidelines on environmental decision-making. However, CONAMA Resolution No. 001/86 does not shed light on how decisions should be made, for instance, which aspects should be considered in an EIA decision. Furthermore, there is no provision on the obligation to make decisions available for public consultation. Therefore, it is necessary to enhance CONAMA Resolution No. 001/86 and include guidelines on decisions as well as procedural obligation to make them available in print in rural areas or places difficult to access and on the internet.

Although administrative decisions must be neutral, according to the Federal Constitution, decisions of major projects are questionable. Massive projects and/or projects where cumulative impacts will be unbearable, and projects planned to be implemented in sensitive environments require special attention since these enterprises are normally controversial and a number of communities are affected. Thus, an independent committee of experts would be important to guarantee neutrality of technical decisions. Furthermore, such a committee would nominate a facilitator to conduct communication with stakeholders.

NOTES

1. Arend J. Kolhoff et al., "The Contribution of Capacities and Context to EIA System Performance and Effectiveness in Developing Countries: Towards a Better Understanding," *Impact Assessment & Project Appraisal* 27, no. 4 (December 2009): 271, 276.
2. Ismo Polonen, "Quality Control and the Substantive Influence of Environmental Impact Assessment in Finland," *Environmental Impact Assessment Review* 26, no. 5 (July 2006): 481, 488.
3. Ibid.
4. International Association for Impact Assessment, "The Linkages between Impact Assessment and the Sustainable Development Agenda, and Recommendations for Actions" (paper, Statements and Policy Briefing for the World Summit on Sustainable Development August 2002), http://www.iaia.org/pdf/IAIAMemberDocuments/News/WSSD/WSSDIAIA10pagerfinal0802.doc.
5. Ibid.
6. Polonen, "Quality Control," 481.
7. See Cristiane Derani, *Direito Ambiental Econômico* (Economic Environmental Law), 3rd ed. (São Paulo: M. Limonad, 2008), 157; Antonio F.G. Beltrão, *Aspectos Jurídicos do Estudo de Impacto Ambiental (EIA)* (Legal Aspects of Environmental Impact Assessment [EIA]) (São Paulo: MP Editora, 2008).
8. Cristiane Derani, *Direito Ambiental Econômico.*
9. Ibid.
10. Raymond M. Robinson, "Environmental Impact Assessment: The Growing Importance of Science in Government Decision Making," *Hydrobiologia* 188, no. 1 (December 1989): 137–142; Angus Morrison-Saunders and John Bailey, "Practitioner Perspectives on the Role of Science in Environmental Impact Assessment," *Environmental Management* 31, no. 6 (June 2003): 683.
11. Derani, *Direito Ambiental Econômico*, 161.
12. Ibid.

13. Francois Retief et al., "Exploring the Psychology of Trade-off Decision-making in Environmental Impact Assessment," *Impact Assessment & Project Appraisal* 31, no. 1 (March 2013): 13.

14. Ibid.

15. Environmental Protection Agency, *Improved Science-based Environmental Stakeholder Processes* (2001), https://yosemite.epa.gov/sab/sabproduct.nsf/cee3f362f1a1344e852571 8e004ea078/$file/eecm01006_report_appna-e.pdf.

16. Robin Gregory and Timothy McDaniels, "Appendix B: Improving Environmental Decision Processes," in Decision Making for the Environment: Social and Behavioral Science Research Priorities, eds. Garry D. Brewer and Paul C. Stern (Washington, DC: National Academies Press, 2005), 190.

17. Ibid.

18. D.N. Tiwari et al., "Environmental Economic Decision-making in Lowland Irrigated Agriculture Using Multi-Criteria Analysis Techniques," *Agricultural Systems* 60, no. 2 (May 1999): 99.

19. Gregory and McDaniels, "Appendix B."

20. UN Conference on Environment and Development, Rio Declaration on Environment and Development, Principle 17, A/CONF.151/26/Rev.1 (Vol. I), Annex I, Principle 15 (Aug. 12, 1992).

21. Gregory and McDaniels, "Appendix B."

22. Ibid.

23. "Criteria for 'Better' Environmental Decision-making," CSERGE—ESRC Programme on Environment, Global Development Research Center, last visited January 6, 2017, https://www.gdrc.org/decision/fs-criteria.html.

24. Rio Declaration on Environment and Development, Annex I.

25. Hens Runhaar et al., "Environmental Assessment in The Netherlands: Effectively Governing Environmental Protection? A Discourse Analysis," *Environmental Impact Assessment Review* 39 (February 2013): 13–25.

26. Netherlands Commission for Environmental Assessment, *New Experiences on EIA in the Netherlands* (1999), http://api.commissiemer.nl/docs/mer/diversen/new_experiences_1998.pdf.

27. Polonen, "Quality Control," 481, 488.

28. Ibid.

29. Ibid.

30. International Finance Corporation, *Policy on Social and Environmental Sustainability*, January 1, 2012, para. 22, https://www.ifc.org/wps/wcm/connect/7141585d-c6fa-490b-a812-2ba87245115b/SP_English_2012.pdf?MOD=AJPERES&CVID=kiIrw0g.

31. Atif Ansar et al., "Should We Build More Large Dams? The Actual Costs of Hydropower Megaproject Development," *Energy Policy* 69 (June 2014): 43–56.

32. UN Human Rights Office of the High Commissioner, *Guiding Principles on Business and Human Rights: Implementing the United Nations "Protect, Respect and Remedy" Framework*, A/HRC/17/31, 2011, 7c, http://www.ohchr.org/Documents/Publications/GuidingPrinciplesBusinessHR_EN.pdf.

33. Robert Goodland, "Free, Prior and Informed Consent and the World Bank Group," *Sustainable Development Law & Policy* 4, no. 2 (2004): 66.

34. Frank Vanclay et al., *Social Impact Assessment: Guidance for Assessing and Managing the Social Impacts of Projects* (2015).

35. Ibid.

36. Ibid.

37. Kolhoff et al., "The Contribution of Capacities and Context," 277.

38. Luis Enrique Sanchez, *Avaliação de Impacto Ambiental: Conceitos e Métodos* (Environmental Impact Assessment Concepts and Methodologies) 2nd ed. (São Paulo: Oficina de Textos 2013), 16.3.

39. Ibid.

40. Ibid.

41. Vanclay et al., *Social Impact Assessment.*

42. Kolhoff et al., "The Contribution of Capacities and Context," 277.

43. Ibid.

44. Sanchez, *Avaliação de Impacto Ambiental: Conceitos e Métodos*, chap. 17.

45. Ibid.

46. Ibid.

47. Angus Morrison-Saunders and Gerard Early, "What Is Necessary to Ensure Natural Justice in Environmental Impact Assessment Decision-making?," *Impact Assessment & Project Appraisal* 26, no. 1 (2008): 29, 30.

48. Substantive decisions are the decisions that consider and reflect the findings of the EIA. See Sanchez, *Avaliação de Impacto Ambiental: Conceitos e Métodos.*

49. See "Procedures," Netherlands Commission for Environmental Assessment, last accessed March 3, 2016, http://www.commissiemer.nl/english/our-services/types-of-advisory-report.

50. Ibid.

51. Sanchez, *Avaliação de Impacto Ambiental: Conceitos e Métodos.*

52. Netherlands Commission for Environmental Assessment, *Keysheet: NCEA's Advice* (September 2011), http://api.commissiemer.nl/docs/cms/FS%2016%20Advisory%20procedure%20NCEA%20ENG%20Final.pdf.

53. Ibid.

54. Ibid.

55. Jos Arts et al., "The Effectiveness of EIA as an Instrument for Environmental Governance: Reflecting on 25 Years of EIA Practice in the Netherlands and the UK," *Journal of Environmental Assessment Policy & Management* 14, no. 4 (December 2012): 25, 30.

56. Impact Assessment Act S.C. 2019, c. 28, s. 1, art. 24–28; 43, https://laws.justice.gc.ca/PDF/I-2.75.pdf.

57. Ibid., art. 36 (1).

58. Ibid., art. 36 (2).

59. Ibid., art. 41 (1).

60. Ibid., art. 37 (2)

61. Impact Assessment Act S.C. 2019, c. 28, s. 1, art. 24–28.

62. International Bank for Reconstruction and Development and International Development Association, *The World Bank Inspection Panel*, Resolution No. IBRD 93-10 (September 22, 1993), https://www.inspectionpanel.org/sites/ip-ms8.extcc.com/files/documents/Resolution1993.pdf.

63. David Hunter et al., *International Environmental Law & Policy*, 5th ed. (St. Paul: Foundation Press, 2015).

64. "About the Inspection Panel," The World Bank, last visited March 29, 2016, http://ewebapps.worldbank.org/apps/ip/Pages/AboutUs.aspx.

65. The Inspection Panel at the World Bank, *Operating Procedures* (2014), https://www.inspectionpanel.org/sites/ip-ms8.extcc.com/files/documents/2014%20Updated%20Operating%20Procedures.pdf.

66. Ibid.

67. Ibid.

68. Hunter et al., *International Environmental Law and Policy*.

69. Ibid.

70. Ibid.

71. Ibid.

72. Ibid.

73. Natalie L. Bridgeman and David Hunter, "Narrowing the Accountability Gap: Toward a New Foreign Investor Accountability Mechanism," *Georgetown International Environmental Law Review* 20 (2008): 187, 213.

74. Rumu Sarkar, "Exercising a Quasi-Judicial Review Through a World Bank Appellate Body," in *International Economic Law: The State and Future of the Discipline*, eds. Collin Picker et al. (Oxford: Hart Publishing, 2008).

75. Ibid.

76. Bridgeman and Hunter, "Narrowing the Accountability Gap," 210.

77. "About the CAO: Who We Are," Compliance Advisor Ombudsman, last visited March 22, 2016, http://www.cao-ombudsman.org/about/whoweare/.

78. Compliance Advisor Ombudsman, *CAO Operational Guidelines* (2013), http://www.cao-ombudsman.org/howwework/documents/CAOOperationalGuidelines2013_ENGLISH.pdf.

79. Ibid.

80. Ibid.

81. For the eligibility criteria, see the *CAO Operational Guidelines*.

82. Ibid.

83. Ibid.

84. Ibid.

85. Ibid.

86. Ibid.

87. Compliance Advisor Ombudsman, *CAO Operational Guidelines*.

88. Bridgeman and Hunter, "Narrowing the Accountability Gap," 210.

89. Ibid.

90. Ibid.

91. "Project Complaint Mechanism," European Bank for Reconstruction and Development, last visited November 5, 2019, https://www.ebrd.com/work-with-us/project-finance/project-complaint-mechanism.html.

92. Ibid.

93. Lei No. 6938, de 31 de Agosto de 1981, Diário Oficial da União (D.O.U.) de 02.09.1981 (Braz.).

94. Sanchez, *Avaliação de Impacto Ambiental: Conceitos e Métodos*, 25.

95. Ibid.

96. Ibid.

97. Sanchez cites one case, in 1994, where the committee voted against the issuance of the permit for a quarry in São Paulo. The author notes that this was the first time where the committee was against the issuance of a permit. See Sanchez, *Avaliação de Impacto Ambiental: Conceitos e Métodos*, 25.

98. Beltrão, *Aspectos Jurídicos*, 81.

99. Ibid.

100. Ibid.

101. Sanchez, *Avaliação de Impacto Ambiental: Conceitos e Métodos*, 25.

102. Ismo Polonen et al., "The Effectiveness of the Finnish EIA System—What Works, What Doesn't, And What Could be Improved?," *Environmental Impact Assessment Review* 31, no. 2 (March 2011): 120, 123.

103. See Eve Bratman, "Passive Revolution in the Green Economy: Activism and the Belo Monte Dam," *International Environmental Agreements: Politics Law & Economics* 15, no. 1 (March 2015): 61.

104. Thomaz Miazaki de Toledo, "Licenciamento Ambiental Federal" (Federal Environmental Licensing) (PowerPoint presentation, Environmental Licensing: Reality and Perspectives, Brasília, November 5, 2015), http://www.mpf.mp.br/atuacao-tematica/ccr4/dados-da-atuacao/eventos/Seminarios/ThomazMiazaki.pdf.

105. Ibid.

106. Ibid.

107. Paulo Affonso Leme Machado, *Direito Ambiental Brasileiro* (Brazilian Environmental Law), 21st ed. (São Paulo: Malheiroso Editores, 2013), 312, 313; Tribunal de Contas da União, *Acórdão No. 2.212 Plenário*.

108. Machado, *Direito Ambiental Brasileiro*, 312, 313.

109. See Bratman, "Passive Revolution in the Green Economy," 61.

110. Fernando Antonio Dorna Magalhães, "Licenciamento Ambiental. Atuação do TCU." (Environmental Licensing. The Role of the Audit Court) (PowerPoint presentation, Environmental Licensing: Reality and Perspectives, Brasília, November 5, 2015), http://www.mpf.mp.br/atuacao-tematica/ccr4/dados-da-atuacao/eventos/Seminarios/Ppt0000053.pdf.

111. Bratman, "Passive Revolution in the Green Economy," 61. Observers foresaw that the government would ignore protests and legal actions and would continue with the implementation of the Belo Monte Dam. Alexei Barrionuevo, "Brazil Rejects Panel's Request to Stop Dam," *New York Times*, April 5, 2011, https://www.nytimes.com/2011/04/06/world/americas/06brazil.html.

112. Cases of corruption involving infrastructure projects have been investigated and in the spotlight in the media. See "Delcidio diz que Dinheiro da Obra de Belo Monte Abasteceu Campanha" (Delcidio States that Money from Belo Monte Implementation Assisted Presidential Campaign), G1, March 12, 2016, 11:03 p.m., http://g1.globo.com/jornal-nacional/noticia/2016/03/delcidio-diz-que-dinheiro-da-obra-de-belo-monte-abasteceu-campanhas.html.

113. Machado, *Direito Ambiental Brasileiro*, 312, 313.

114. Ibid., 313.

115. See Peter Taylor Klein, "Engaging the Brazilian State: The Belo Monte Dam and the Struggle for Political Voice," *Journal of Peasant Studies* 42, no. 6 (March 2015): 1137–156.

116. See generally Ibid., 1137–1156.

117. TRF-1, A.I. No. 11749-24.2015.4.010000/Pa Pará, Relator: Des. Jirair Aram Megueriam, 24.04.2015 (Braz.).

118. Ibid.

119. Public Hearing carried out on Jan. 10, 2013. Transcription in the Environmental Public Procedure No. 5028/2012.

120. Interviews with anonymous citizens and public officials involved with the environmental permitting of the Volta Grande Project, in Altamira, Senador José Porfírio, and Belém, Pará State, Brazil, July 2–9, 2015.

121. Ibid.

122. Ibid.

123. *See* Beltrão, *Aspectos Jurídicos*.

124. Council on Environmental Quality (CEQ) Regulations for Implementing the Procedural Provisions of the National Environmental Policy Act, 40 C.F.R. § 1505.2 (2015).

125. Constituição Federal (C.F.) (Constitution) art. 37 (Braz.).

126. Ibid.; Lei No. 9.784, de 29 de Janeiro de 1989, Col. Leis Rep. Fed. Brasil, 1, 41, Fevereiro 1999 (Braz.).

127. Council on Environmental Quality (CEQ) Regulations for Implementing the Procedural Provisions of the National Environmental Policy Act, 40 C.F.R. pts. 1500-1508 (2012).

128. Ibid., Sec. 1506.1 Limitations on actions during NEPA process. "Until an agency issues a record of decision as provided in Sec. 1505.2 (except as provided in paragraph (c) of this section), no action concerning the proposal shall be taken which would: 1. Have an adverse environmental impact; or 2. Limit the choice of reasonable alternatives."

129. Instrução Normativa do IBAMA No. 184, de 17 de Julho de 2008, Diário Oficial da União (D.O.U.) 18.07.2008 (Braz.). It was amended by Instrucao Normativa do IBAMA No. 14/2011.

130. See "Sistema Informatizado de Licenciamento Ambiental" (Online System of Environmental Licensing), IBAMA (last accessed December 8, 2019), http://www.ibama.gov.br/sistemas/sislic/sobre-o-sislic.

131. Instituto Socioambiental and Rainforest Foundation Norway, *Consulta Livre, Prévia e Informada na Convenção 169 da OIT* (Free, Prior and Informed Consent in Light of the ILO Convention No. 169) (undated), https://site-antigo.socioambiental.org/inst/esp/consulta_previa/?q=convencao-169-da-oit-no-brasil#quem.

132. Decreto No. 143, de 20 de Junho de 2002, Diário Oficial da União (D.O.U.) de 21.06.2002 (Braz.).

133. International Labour Organization (ILO), Indigenous and Tribal Peoples Convention, art. 1, C169, 27 June 1989, C169.

134. Quilombolas are afrodescendents from traditional communities who, during slavery, ran away from their "owners" to quilombos. Therefore, communities who live in Quilombos are called quilombolas.

135. Constituição Federal (C.F.) (Constitution) art. 231 (Braz.).
136. Ibid, art.68.
137. Instituto Socioambiental and Rainforest Foundation Norway, *Consulta Livre*.
138. Ibid.
139. J.F. 5a São Luis do Maranhão, Mandado de Segurança No. 2006.37.00.005222-7, São Luis do Maranhão 13.02.2007 (Braz.).
140. Instituto Socioambiental and Rainforest Foundation Norway, *Consulta Livre*.
141. *Saramaka People v. Suriname*, Preliminary Objections, Merits, Reparations, and Costs, Judgment, Inter-Am. Ct. H.R. (Ser. C) No. 185 (November 28, 2007).
142. Instituto Socioambiental and Rainforest Foundation Norway, *Consulta Livre*.
143. Deborah Duprat, "A Convenção n. 169 da OIT e o Direito a Consulta Livre, Prévia e Informada" (The ILO Convention No. 169 and the Free, Prior and Informed Consent), *in Convenção No. 169 da OIT e os Estados Nacionais* (ILO Convention No. 169) (Deborah Duprat ed., 2015), http://www.mpf.mp.br/atuacao-tematica/ccr6/documentos-e-publicacoes/artigos/docs_artigos/convencao-169-da-oit_web.pdf.
144. "Povos e Comunidades Tradicionais do Brasil" (New Social Cartography of the Amazon, Peoples and Traditional Communities of Brazi), Nova Cartografia Social da Amazônia, last visited November 30, 2016, http://novacartografiasocial.com/fasciculos/povos-e-comunidades-tradicionais-do-brasil/?cp=2.
145. Ibid.
146. Decreto 6.040, de 7 de Fevereiro de 2007, Diário Oficial da União (D.O.U.) de 08.02.2007 (Braz.).
147. Duprat, "A Convenção n. 169 da OIT e o Direito a Consulta Livre."
148. Ibid.
149. Ibid.
150. Instituto Socioambiental and Rainforest Foundation Norway, *Consulta Livre*.
151. Duprat, "A Convenção n. 169 da OIT e o Direito a Consulta Livre."
152. J.F. 1, Ação Civil Pública (ACP) No. 0006962-86.2014.4.013200, Amazonas, 23.05.2014 (Braz.).
153. Indigenous and Tribal Peoples Convention, art. 1(2). Self-identification as indigenous or tribal shall be regarded as a fundamental criterion for determining groups to which the provisions of this convention apply.
154. See Thiago Felipe S. Avanci, "Collision of Fundamental Rights: Is There Collision of Fundamental Rights?," *Revista Brasileira de Direito Constitucional* 16 (November 2010): 193, http://www.esdc.com.br/RBDC/RBDC-16/RBDC-16-193-Artigo_Thiago_Felipe_S._Avanci_(A_Colisao_de_Direitos_Fundamentais).pdf. See also Alexandre de Castro Coura and Lívia Avance Rocha, "Tutela Jurisdicional do Meio Ambiente, e Fenômeno da Colisão de Direitos Fundamentais na Jurisprudência do STF e do STJ" (Jurisdictional Protection of the Environment, and Phenomenon of Collision of Fundamental Rights in the Jurisprudence of the Supreme Court and the Superior Tribunal of Justice), *Revista de Direito Ambiental* 19, no. 76 (2014): 37–60 (analyzing Brazilian jurisprudence in light of Robert Alexys theory of balance test of fundamental rights).

155. Art. 3, II of the Federal Constitution states that "The fundamental objectives of the Federative Republic of Brazil are: II—to guarantee national development." Constituição Federal (C.F.) (Constitution) art. 3, II (Braz.).

156. Celso Antonio Pacheco Fiorillo et al., *Licenciamento Ambiental* (Environmental Licensing) (Brazil: Saraiva, 2011): 27.

157. Goodland, "Free, Prior and Informed Consent and the World Bank Group," 66.

158. Instituto Socioambiental & Rainforest Foundation Norway, *Consulta Livre*.

159. Ibid.

160. Duprat, "A Convenção No. 169 da OIT e o Direito a Consulta Livre," 53.

161. Ibid.

162. Ibid.

163. Ibid.

164. Maria Luisa Grabner, "O Direito Humano ao Consentimento Livre, Prévio e Informado como Baluarte do Sistema Jurídico de Proteção dos Conhecimentos Tradicionais" (The Human Right to Free, Prior and Informed Consent as Bulwark of the Judicial System to Protect Traditional Knowledge), *Boletim Científico ESMPU* 45, no. 14 (July 2015): 29.

165. Ibid.

166. Instituto Socioambiental & Rainforest Foundation Norway, *Consulta Livre*.

167. International Labor Organization, *Handbook for ILO Tripartite. Understanding the Indigenous and Tribal Peoples Convention, 1989 (ILO 169)* (2013), https://www.ilo.org/wcmsp5/groups/public/---ed_norm/---normes/documents/publication/wcms_205225.pdf.

168. Duprat, "A Convenção No. 169 da OIT e o Direito a Consulta Livre," 53.

169. Ibid.

170. Environmental Protection Agency, Improved Science-based Environmental Stakeholder Processes.

171. Instituto Socioambiental and Rainforest Foundation Norway, *Consulta Livre*.

172. Ibid.

173. Ibid.

174. Ibid.

175. Impact Assessment Act S.C. 2019, c. 28, s. 1, art. 75 (1), https://laws.justice.gc.ca/PDF/I-2.75.pdf.

176. Instituto Socioambiental & Rainforest Foundation Norway, *Consulta Livre*.

177. Ibid.

178. Prosecutor Thais Santi, during a public hearing on September 13, 2012, to discuss the Volta Grande Project, stated, "I consider the communities who live in the area of the project as traditional communities, riverbank [communities]. Therefore, they cannot be relocated as we move furniture." Licenciamento Ambiental do Projeto Volta Grande (Volta Grande Environmental Permitting Procedure) No. 5028/2012 (on file with author).

179. Interviews with anonymous citizens and public officials involved with the environmental permitting of the Volta Grande Project.

180. Interviews with anonymous citizens and public officials involved with the environmental permitting of the Volta Grande Project.

181. L. Ortolano, "Controls on Project Proponents and Environmental Impact Assessment Effectiveness," *Environmental Professional* 15, no. 4 (January 1993): 352–363.

182. Y. Destra et al., "Using Knowledge for Decision-making Purposes in the Context of Large Projects in the Netherlands," *Environmental Impact Assessment Review* 23, no. 5 (August 2003): 517, 521.

183. Ibid.

184. "Política de Responsibilidade Social e Ambiental" (National Development Bank, Social and Environmental Policy), Banco Nacional de Desenvolvimento Econômico e Social (BNDES), last visited November 5, 2019, http://www.bndes.gov.br/wps/portal/site/home/quem-somos/responsabilidade-social-e-ambiental/o-que-nos-orienta/politicas/politica-responsabilidade-social-ambiental.

185. Alan Bond et al., "Progressing Quality Control in Environmental Impact Assessment Beyond Legislative Compliance: An Evaluation of the IEMA EIA Quality Mark Certification Scheme," *Environmental Impact Assessment Review* 63 (March 2017): 160.

186. James L. Creighton et al., *Involving Citizens in Community Decision Making: A Guidebook* (Washington, DC: Program for Community Problem Solving, 1994).

187. Ibid.

188. In 2009, the Brazilian Auditing Court recommended the creation of coordination at IBAMA responsible exclusively for reviewing Environmental Impact Statements and communicating its results. See Tribunal de Contas da União, *Acórdão No. 2.212 Plenário, Relatório de Levantamento. Avaliação do IBAMA* (Report of Analysis of the Federal Environmental Agency—IBAMA) (September 23, 2009), https://pesquisa.apps.tcu.gov.br/#/resultado/acordao-completo/*/NUMACORDAO%253A2212%2520ANOACORDAO%253A2009/%2520.

189. In Canada, the Independent Environmental Monitoring Agency, a watchdog committee, was created to monitor the impacts of the Ekati diamond mine. The agreement, which established the independent agency, stated that members of the agency would include a retired senior politician from the government of the Northeast Territories, a university professor specialized in wildlife, and an environmental consultant with impact assessment experience, mostly working for aboriginal consultants, among others. See William A. Ross, "The Independent Environmental Watchdog: A Canadian Experiment in EIA Follow-up," in *Assessing Impact: Handbook of EIA and SEA Follow-up*, eds. Angus Morrison-Saunders and Jos Arts (London: Routledge, 2004); Information about the Independent Environmental Monitoring Agency is available in its website at, Independent Environmental Monitoring Agency (last accessed December 8, 2019), https://monitoringagency.net/about/#the-agency.

190. Kolhoff et al., "The Contribution of Capacities and Context," 277.

191. As Law No. 7.735 of February 22, 1989, created IBAMA and stated its responsibilities, a federal law would be necessary to create the EIA independent committee.

192. Some scholars understand that environmental permitting procedures should be regulated by law rather than by a CONAMA Resolution. However, in practice the type of norm, whether it is a law or a CONAMA Resolution does not change the state of the art. CONAMA Resolution No. 001/86 has been applied and observed, at least in theory, for the last thirty years in Brazil.

Conclusion

EIA procedures are recognized as a practice necessary to aid environmental and economic decision-making. Public participation is an important step in EIA procedures worldwide. However, it is not defined under international law when and how much weight participation should have in projects' decision-making.

Under the UNDRIP and ILO Convention No. 169, indigenous and tribal peoples should provide FPIC before authorization of project is given. In practice, it has been difficult to enforce this right, because national governments and development institutions are not willing to truly discuss and reach consensus with these communities. Nonetheless, indigenous peoples are achieving new heights in the development debate. For example, the IACHR has affirmed the right of indigenous peoples to FPIC. This book introduces the concepts of consultation, FPIC, and consent as a background to develop an EIA procedure to reach eventual consensus of local affected communities.

Looking at the Brazilian case, the Brazilian Federal Constitution determines that EIA must be prior to authorization of projects and shall be made public. The EIA is not a formal procedure per se. It happens within the environmental permitting, which is comprised of three stages: preconstruction, installation, and operation permits. The realization of an EIA is a condition to the issuance of the preconstruction permit. EIA is required for activities listed in CONAMA Resolution No. 001/86—the main EIA regulation at the federal level.

One example to understand how EIA legislation actually works in the Brazilian context is the Volta Grande Project, an open-pit mine slated to be implemented in a very sensitive region, in the Amazon rainforest: the Volta Grande area, which is a 100-kilometers curve of the Xingu River, is home to ornamental fishes. The project is also located at the center of three indigenous lands. The area of the project has been inhabited by artisanal miners for decades. Other communities, such as riverbank residents and fishermen, would potentially be affected if the mining activity is authorized by the state environmental agency. In summary, this case study is interesting because it involves different social groups.

Aside from these natural aspects, the Volta Grande is also the area of influence of the Belo Monte Dam. The dam started filling its reservoir for operation in 2016, yet its environmental and social impacts began much earlier, especially in the Volta Grande area, during the construction of the dam. Residents reported a decrease in fish stocks, which is the main source of protein in their diet. Considering that the normal flow of the Volta Grande would be diverted to generate energy in Belo Monte's power house, the quality and amount of water in this specific portion of the river ought to be monitored by Belo Monte for a period of six years. Therefore, there was uncertainty on how nature would react during operation of the dam.

In this fragile and already impacted environment, the Volta Grande Project has been engaged in the environmental permitting procedure, aiming for the installation permit to enable the start of construction. The preconstruction permit was issued in 2014, despite public concerns and legal actions questioning several aspects of the EIA and the environmental permitting. This includes ineffective public participation, competence of the state environmental agency to issue the permit, lack of consultation of indigenous communities, and lack of proper assessment of cumulative impacts of both mining and damming. In 2018, the project's permit was suspended by the federal court on a decision determining that the environmental permitting be transferred to the federal environmental agency.

This case demonstrates in practice the weaknesses of public participation in Brazilian EIAs. First, the EIA procedure was not transparent and access to information was flawed. The EIS was not easily available for local affected communities. Second, the first public hearing to explain the project to the population was held far from where local affected community lived and the language used was extremely technical.

The Volta Grande also reveals a lack of coordination among governmental agencies which should have been involved at early stages of the process due to the localization of the project on public lands, surrounded by indigenous communities, and with a high incidence of malaria. Therefore, all agencies responsible for these issues should have participated in the beginning of the EIA. This practical example provides interesting aspects for a theoretical consideration regarding EIA in a developing context.

Practitioners raise the argument that Brazilian environmental permitting procedures are already long due to overburdened legal actions; therefore, measures to include more public participation would make this procedure even longer and eventually more expensive. Research demonstrates that EISs are fraught with poor quality regarding data collection, lack of assessment of cumulative impacts, among others. As a consequence, these assessments are questioned in court. Moreover, lack of appropriate public participation is frequently questioned by the *Ministério Público*. Increased stakeholders' participation would make the process more legitimate and increase social control, which could avoid poor quality studies; consequently, reducing the chances to be questioned in court. In other words, the actual Brazilian environmental permitting procedures are considered long and complex, but the lack of public participation is actually one of the reasons of its complexity and time consumed. It is also more complicated to have political maneuvers trying to push a project on the population when the EIA and decision-making are transparent and participatory.

Levels of public participation are divided into four categories: informing, consulting, participating, and negotiating. Normally, EIAs require consultation, which is a two-way communication between stakeholders, governments, and project proponents. Therefore, people have the right to be listened to and their views should be taken into consideration in the decision of projects' approval. There are methods to ensure consultation, such as written comments period, workshops, and public hearings. Transparency and access to information are essential to ensure meaningful consultation.

The benefits of meaningful public participation throughout the EIA procedure are crucial. Moreover, it is key that it starts at early stages to inform and educate communities, increasing their chances to influence decisions. Therefore, the question of who should decide on approvals of projects is of utmost importance since political will supersedes social and environmental concerns, even though in some cases there is strong public participation in EIAs.

Consultation only, depending on the project and certain circumstances, is not enough. A process that enables meaningful participation and ensures consensus of affected communities is deemed necessary. Especially in developing countries where the benefits of economic growth often do not reach those who are most affected by negative impacts of projects. These communities sometimes are displaced from their land, lose their social and physical ties with their previous lives; or when they are not displaced, certain neighboring activities can be a constant nightmare due to noise, an increase in traffic, pressure on public services (hospitals, public transportation, etc.). One could say that these projects improve people's lives, bring jobs, improve cities' infrastructure. Although this sounds theoretically plausible, reality generally provides a completely different picture. This book proposes measures to avoid this scenario of development with significant unwanted consequences.

In the Brazilian context, the installation of Belo Monte has not yet brought the promised benefits, especially to those who live in Altamira, the most affected city by this dam. In fact, people's discontentment with high levels of criminality, prostitution, rents' prices, food, and transportation remains extremely relevant in a city lacking public infrastructure. For example, in 2015, it was observed open sewage flowing on the streets as well as mountains of trash on sidewalks attracting many vultures. Moreover, the impacts are extended to indigenous and rural communities located in the area of influence of the dam, such as SJP.

Therefore, FPIC should be extended to all local affected communities. It is a matter of justice and democracy to account for their consensus in relation to decisions affecting their lives. It is necessary to clarify that this proposal does not refer to veto power, but as a necessity to have meaningful dialogues and a serious compromise to construct a common decision. Nonetheless, providing a more democratic decision-making in EIA procedures benefits also those interested in investing in large projects in any area, especially in fragile ecosystems as the Amazon. Clear and transparent rules for public participation in EIA provide a business environment that fosters investments, which might benefit all economic agents, including those in affected areas. ILO Convention No. 169 recognizes FPIC requirements for indigenous and tribal peoples, although FPIC does not mean that these communities can veto a project. However, government decisions cannot be imposed on them. In fact, unilateral decisions should not be imposed on each other, either on the part of the government or the communities.

Local affected communities under specific circumstances should have the right to FPIC, as a process to achieve their consensus or quasi-consensus. This proposal does not undermine in any manner indigenous communities' right to FPIC. The idea is to expand this concept to local communities who are affected by large-scale projects or by cumulative impacts of several activities. It is necessary to have a process of real dialogue among government, project proponents, intervening agencies, and communities.

Currently, public participation within Brazilian EIAs is not effective. Generally, it is carried out in order to comply with formal requirements. Stakeholders are informed of an EIS and react, many times, negatively to it. This is not public participation and it is not a two-way means of communication. Furthermore, there is no guarantee that communities' input will be taken into account at the final decision, which is not in many cases. Public hearings alone are not enough to guarantee meaningful participation. For these and other reasons indicated throughout this book, an EIA should be not only a technical study but also a procedure to engage, educate, and consult stakeholders to reach their consensus.

In answering the argument that the voice of a minority would delay development and, therefore, the interest of the rest of the country, affected communities' right to life, housing, and a clean environment is not less important than the right of the country to development. Disasters, such as those occurred in Mariana in 2015 and Brumadinho in 2019, are examples that local communities are the most affected by any error in mining projects. Local communities' views should be truly considered in an EIA and ultimately in the decision authorizing a project, since their lives and properties are under risk as a consequence of a large project such as mining. Moreover, local communities' acceptance of projects is very beneficial for business because it avoids strikes, protests, delays in environmental permitting procedures as well as affecting the private sector's image. The case study analyzed is a classic example. Belo Sun stated in an appeal at a public civil action that every time there was news circulated in the media of delays in its environmental permitting procedure, there was a financial impact on the project. In their own words, "Delay in the environmental permitting procedure and the actions of the *Ministério Público* against the project is one of the reasons that result in fall of their share prices. Belo Sun's share dropped from CAD 1.80 to lower than CAD 0.20 [in 2013]."[1]

A major issue in Brazil as well as in other developing countries is how to establish criteria capable of diminishing governments' discretion when projects raise serious social and environmental concerns. How to prevent political and economic decisions from overriding social and environmental protection is a question still to be further developed. Decision-making literature acknowledges that criteria for administrative decisions are an area which requires in-depth research. While EIA legislations are procedural by nature, and Brazil is not an exception, this book briefly looks at how substantive criteria could be applied in reviewing EIA by an independent committee as well as for decision-makers in the authorization of projects.

To ensure impartiality and engagement of communities and other stakeholders, an independent committee would be in charge of conducting dialogues and reviewing the EIA. Legislation would guide this committee in its duty to review the EIA. The proposed independent committee would have two functions: one in relation to stakeholder's engagement and another function reviewing technical quality of the assessment.

This measure would separate the technical from the political decision on approval of projects. The independent committee would not be in charge of issuing the permit, which could remain with the environmental agency. However, the environmental agency would have to comply with substantive criteria when issuing its decision. For instance, the agency would need to analyze and answer in writing the following questions: Will the project affect people's human right to life? Is this project technology, location, and design the most social environmentally friendly alternative? Having these substantive criteria established in the legislation gives leverage to challenge noncompliant decisions in court. Furthermore, decisions must be available to public consultation through environmental agencies' website and other means of communication.

To summarize, this book proposes a full public participation EIA procedure. It proposes measures to support public engagement during the project life cycle and requires feasibility studies for certain projects—a practice that is already being carried out voluntarily by large companies in Brazil. The screening would select those projects which are large-scale, have severe potential impacts, have significant cumulative impacts, and/or are slated to be implemented in delicate environments. When the project falls under one or more of these criteria, it would require FPIC of local affected communities. A compromise of discussions and dialogue would be carried out through scoping workshops to assist the development of the TOR for the EIA. Once the TOR has been written, a public hearing would be important to present the final document to communities. Subsequently, once the EIS is completed a defined minimum period of at least thirty days would be open to receive written comments. This period would be followed by obligatory public hearings to discuss the EIA and its findings. All these steps should be followed under the review of an independent committee.

In the long term, implementation of the procedure proposed here would bring enormous benefits to EIAs and promote social and environmentally sound decisions on development projects. Table C.1 summarizes all the changes and the flowchart in figure C.1 represents the EIA procedure proposed in this study.

In recent years, Brazil went through some tough moments. Two dreadful mining accidents marked the period 2015–2019, both in southeast of Minas Gerais. In November 2015, the bursting of a mining sludge dam in the city of Mariana caused irreversible losses to both human lives and the environment.[2] The sludge flooded over 900 kilometers contaminating the water all the way to the sea in the state of Espírito Santo, killing nineteen people and destroying houses and natural resources. Again, in January 2019, a dam with contaminated mud collapsed in Brumadinho. The mudslide buried everything on its way killing around 300 people.[3]

Recalling the dialogue in one of the public hearings of the Belo Sun project, local affected communities enquired about the use of explosives on the rocks to extract gold and whether these explosions would affect the Belo Monte Dam, resulting in its burst. The project proponent answered vaguely, using highly technical terms stating that this risk was still under study; however, the company believed this possibility was very low. Moreover, the environmental agency representative explained that there was a difference between risks and impacts and that the referred study would be on the risk of both projects, giving the impression that the objective of the public hearing was to explain just the impacts and not the risks of the projects. The public hearing of the Volta Grande Project was held way earlier than these incidents involving mining activities in Brazil, yet concerns raised by local communities were valid and seemed appropriate to be discussed in that forum. Their fear should also have been managed and addressed properly, but unfortunately this did not happen.

Unfortunately, these events corroborate that local affected communities are those who are heard but not listened to. Although project proponents and environmental agencies have in place a "consultative" process, by carrying

Table C.1 Proposal to Improve Public Participation in Brazil

Phase of EIA Procedure	*Proposal*
1. Planning the Project	Project proponent carries out a social and environmental feasibility study.
2. Screening	Two lists of projects would define whether the project would require full public participation EIA or simpler public participation EIA procedure.
3. Mapping Stakeholders	Project proponent with the assistance of the independent committee would identify stakeholders to participate in the full public participation EIA procedure.
4. Scoping	Before drafting the TOR for the EIA. The independent committee would carry out the scoping workshops. The committee would ensure that a neutral facilitator conduct the workshops and accessible language be used.
5. Write the TOR	The project proponent writes the TOR considering input collected during scoping workshops.
6. Public Hearing on the TOR	The EIA independent committee holds a public hearing to present the TOR to stakeholders.
7. Draft the EIS	The EIS must be made timely available prior to consultation. Alternative means of communicating written comments and public hearings must be created, such as project websites, e-mail lists, advertising on social media, radio, TV, loudspeakers.
8. Written Comments/ Public Hearing	The independent committee ensures that a reasonable time be open to collect comments from stakeholders and a public hearing be held by a neutral facilitator.
9. Review the EIS	The independent committee will review the EIS and ensure that a consensus or quasi-consensus be reached (FPIC). The approval will consider whether public participation was meaningful and appropriate, whether the data presented and methodologies were suitable and how the people and the environment will be impacted by the project.
10. Issuance of the Environmental Permit	The environmental agency will issue the permit based on the criteria established in the law.

* Voluntary Action

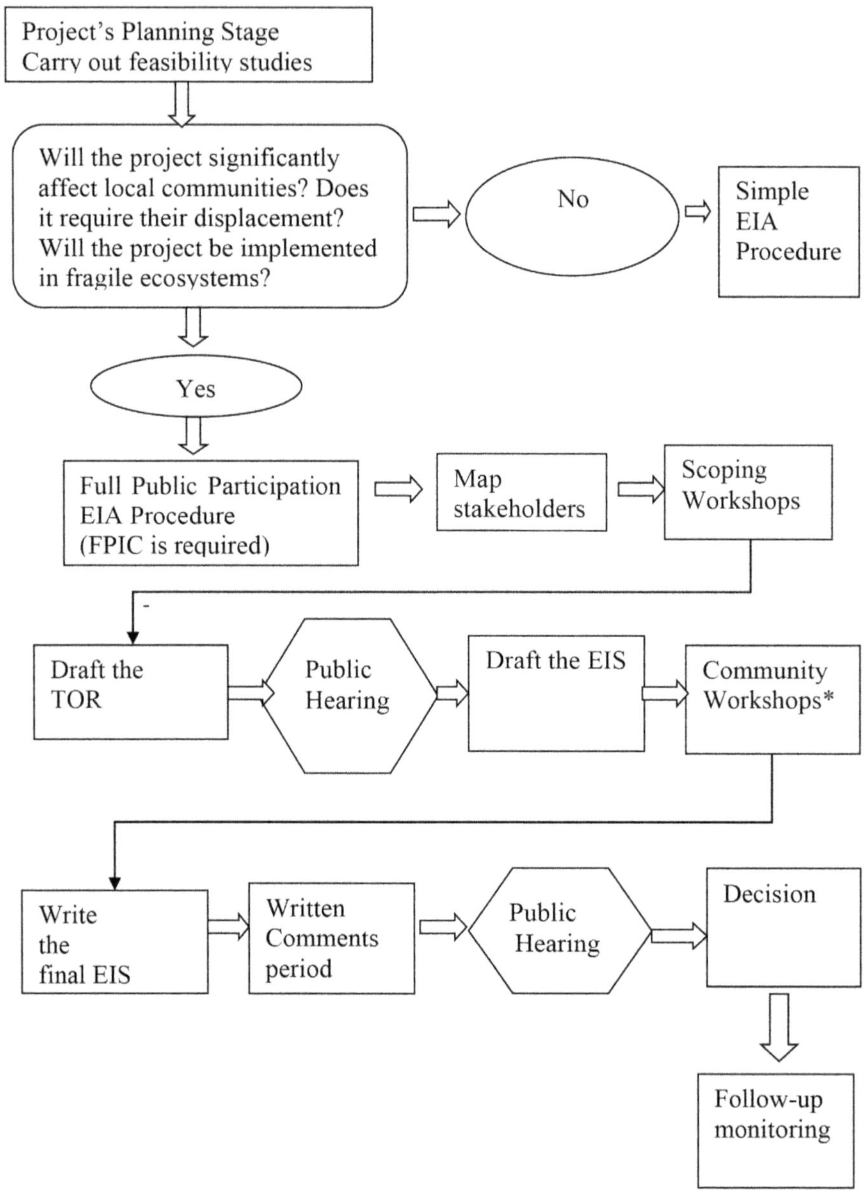

Figure C.1 Proposal: Flowchart of EIA Procedure in Brazil.

out public hearings to inform the public about projects' impacts, therefore complying with the legislative requirement, people's views are not seriously taken into account. Ultimately these same people are those who suffer the consequences of poor quality EIAs and therefore approval of harmful projects.

As a result of these events, what is seen in Brazil is twofold. An increasing distrust in these methods of "participation" emerges, and on the other hand, the government intent on rushing environmental permitting processes, including eliminating the realization of public hearings all together. In 2019, during a public hearing to present the impacts of a mining activity in Minas Gerais, one attendee gave an emotional speech against environmental harmful projects and his sorrow related to the victims of the Brumadinho accident. The attendee physically assaulted project proponents and a public servant who were presenting in the hearing. Unfortunately, this scenario is a result of lack of real community engagement.[4]

Instead of increasing public participation, Brazil is looking for ways to decrease it. Draft laws are proposed by the senate proposing a simpler and faster environmental permitting procedure for strategic projects, which are of national interest.[5] This includes exempting the necessity of public hearings during the EIA procedure.

It is necessary to revise the objectives of the country and examine what kind of development Brazil wants to adopt. In this context, this book adopts the "do-it-right" approach. Simply put, effective public participation leading to FPIC in EIA procedure might take some time, but the social and financial benefits will certainly be worthwhile. Sustainable development can be achieved if people are actually listened to and not be only vaguely heard.

NOTES

1. TRF-1 A.I. No.11749-24.2015.4.010000/Pa Pará, Relator: Des. Jirair Aram Megueriam, 24.04.2015 (Braz.).

2. See G1, *PF Conclui Inquérito da Tragédia de Mariana e Indicia 8 Pessoas* (Federal Police Concludes Investigation on the Tragedy that happened at Mariana and Indicts 8 people) (June 9, 2016, at 9:03 PM), http://g1.globo.com/minas-gerais/desastre-ambiental-em-mariana/noticia/2016/06/pf-conclui-inquerito-da-tragedia-de-mariana-e-indicia-8-pessoas.html.

3. BBC News Night, Brazil dam collapse: a disaster waiting to happen?, https://www.bbc.co.uk/programmes/p0783tk5; Brazil dam: how German firm approved Brazil dam before it burst, https://www.bbc.com/news/world-europe-48948775.

4. G1, Minas Gerais, Jovem agride participantes de audiência pública sobre projeto da Vale em Catas Altas; Veja vídeo (A Young man attacks participants at a public hearing about Vale's project at Catas Altas; Watch the video) https://g1.globo.com

/mg/minas-gerais/noticia/2019/12/18/jovem-agride-participantes-de-audiencia-publica-sobre-projeto-da-vale-em-catas-altas-veja-video.ghtml.

5. The draft law No. 654/2015 proposes much faster environmental permitting procedure to projects considered of government interest. In total the environmental permitting procedure would last eight months. See Projeto de Lei do Senado No. 654 de 25 de Novembro de 2015, https://www25.senado.leg.br/web/atividade/materias/-/materia/123372 *See also* Alberto Fonseca et al., "Reforming EIA Systems: A Critical Review of Proposals in Brazil," *Environmental Impact Assessment Review* 62 (2017); 90, 96.

Bibliography

Affonso Leme Machado, Paulo. "Legislação de Avaliação de Impactos Ambientais no Brasil" [Legislation on Environmental Impact Assessment in Brazil], *Manual de Avaliação de Impactos Ambientais* [Environmental Impact Assessment Guide], edited by Secretaria Especial do Meio Ambiente do Paraná (SEMA-PR). Curitiba: Surehma, 1992, as cited in Oliveira, Antonio Inage de Assis Oliveira, *O Licenciamento Ambiental* [Environmental Licensing]. Brazil: Iglu, 1998.

Amaral Rocha, Tiago and Mariana Oliveira Barreiros de Queiroz. "O Meio Ambiente como Direito Fundamental da Pessoa Humana" [The Environment as a Fundamental Right], *Revista Âmbito Jurídico* 95 (Dec. 2011). http://ambito-juridico.com.br/site/?artigo_id=10795&n_link=revista_artigos_leitura.

"Amazon Fires: Brazil Threatened over EU Trade Deal." *BBC News*, August 23, 2019, https://www.bbc.com/news/world-latin-america-49450495/.

American Convention on Human Rights, November 21, 1969, 1144 U.N.T.S. 143.

Ansar, Atif, Bent Flyvbjerg, Alexander Budzier, and Daniel Lunn. "Should We Build More Large Dams? The Actual Costs of Hydropower Megaproject Development." *Energy Policy* 69 (June 2014): 43–56. https://doi.org/10.1016/j.enpol.2013.10.069.

Appah-Sampong, Ebenezer. *Public hearing within the environmental impact assessment review process*. UNEP Environmental Impact Assessment Training Resource Manual. Last accessed February 22, 2020. http://www.unep.ch/etu/publications/16%29%2085%20to%2091.pdf.

Arts, Jos, Hens A. C. Runhaar, Thomas B. Fischer, Urmila Jha-Thakur, Frank Van Laerhoven, Peter P. J. Driessen and Vincent Onyango. "The Effectiveness of EIA as an Instrument for Environmental Governance: Reflecting on 25 Years of EIA Practice in the Netherlands and the UK." *Journal of Environmental Assessment Policy and Management* (JEAPM), World Scientific Publishing Co. Pte. Ltd. 14, n. 04 (Dec. 2012): 25.

Assunção, Francisca Neta Andrade, Maria Augusta Almeida Bursztyn, and Teresa Lúcia Muricy de Abreu. "Participação social na avaliação de impacto ambiental: lições da experiência da Bahia." [Social Participation in Environmental Impact

Assessment: Lessons from the Experience in Bahia], *Revista Franco Brasileira de Geografia* 10, n. 10 (2010). https://confins.revues.org/6750.

Avanci, Thiago Felipe S. "A Colisão de Direitos Fundamentais: Há Colisão de Direitos Fundamentais?" [Collision of Fundamental Rights: Is There Collision of Fundamental Rights?] *Revista Brasileira de Direito Constitucional* n. 16 (Jul./Dec. 2010): 193–215. http://www.esdc.com.br/RBDC/RBDC-16/RBDC-16-193-Artigo_Thiago_Felipe_S._Avanci_(A_Colisao_de_Direitos_Fundamentais).pdf.

Avritzer, Leonardo. *Participatory Institutions in Democratic Brazil*. Woodrow Wilson Center Press, Washington, DC, 2009.

Azevedo de Lima, Lucas and Simone Amalia Calili. "Uma Análise sobre a Responsabilidade do Conselho Estadual de Determinar a Tipologia para o Município Promover o Licenciamento das Atividades de Impacto Ambiental" [An Analysis of the Responsibility of the State Council to Determine the Competence of the Municipality to Conduct Environmental Licensing of Activities that Causes Environmental Impacts], *Interfaces Científicas* 1 n. 3 (2013): 53–70. http://dx.doi.org/10.17564/2316-381X.2013v1n3p53-70.

Badr, El-Sayed A. "Evaluation of the Environmental Impact Assessment System in Egypt." *Impact Assessment & Project Appraisal* 27, n. 3 (2009): 193–203. 10.3152/146155109X465959.

Baker, Roozbeh B. "Customary International Law in the 21st Century: Old Challenges and New Debate." *European Journal of International Law* 21, n. 1 (May 2010): 173–204. http://ejil.org/pdfs/21/1/1988.pdf.

Baptista, Rodrigo "Avança Projeto que Acelera Liberação de Licenças Ambientais para 'Obras Estratégicas'" [Proposal of Granting Faster Environmental Licenses Authorizing Implementation of Governmental Strategic Projects Moves Forward], *Agência Senado*, November 26, 2015 10:26 a.m., http://www12.senado.leg.br/noticias/materias/2015/11/25/avanca-projeto-que-acelera-liberacao-de-licencas-ambientais-para-obras-estrategicas.

Barelli, Mauro "Free, Prior and Informed Consent in the Aftermath of the UN Declaration on the Rights of Indigenous Peoples: Developments and Challenges Ahead." *International Journal of Human Rights*, 16, n. 1 (Nov. 2012): 1–24. https://papers.ssrn.com/sol3/papers.cfm?abstract_id=1991731.

Barrionuevo, Alexei. "Brazil Rejects Panel's Request to Stop Dam." *New York Times*, April 5, 2011. https://www.nytimes.com/2011/04/06/world/americas/06brazil.html.

Barone, Daniela Marcos. "A Proteção Internacional do Segredo Industrial" [The International Protection of the Industrial Secret] master's thesis, Faculdade de Direito da USP, 2009. http://www.teses.usp.br/teses/disponiveis/2/2135/tde-19112009-133733/pt-br.php.

Barros Vasconcelos, Lucivaldo. "O Estado (In) Transparente: Limites do Direito a Informação Socioambiental no Brasil," [The Lack of Transparency of the State: Limits of Social and Environmental Information in Brazil] PhD diss., Universidade de Brasília, 2008. http://repositorio.unb.br/handle/10482/4167?mode=full.

Beanlands, Gordon E., and Peter N. Duinker. *An ecological framework for environmental impact assessment in Canada*. Halifax, N.S.: Institute for Resource and Environmental Studies, Dalhousie University and Federal Environmental Assessment Review Office, 1983.

Beattie, Robert B. "Everything you Already Know about EIA (But Don't Often Admit)." *Environmental Impact Assessment Review* 15, n. 2 (March 1995): 109–114. https://doi.org/10.1016/0195-9255(95)00001-U.

Becker, Bertha K. and Claudio A. G. Egler. *Brazil: A New Regional Power in the World Economy.* Cambridge: University Press, 1992.

"Belo Monte Usina Hidrelétrica," Agência Nacional de Energia Elétrica, last accessed November 27, 2019. http://www2.aneel.gov.br/aplicacoes/hotsite_beloMonte/index.cfm?p=7.

"Belo Sun," Instituto Socioambiental, last accessed January 14, 2017. https://www.socioambiental.org/pt-br/tags/belo-sun.

Belo Sun Mining, "About Us," last accessed November 23, 2019. https://www.belosun.com/about_us/overview/.

Beltrão, Antônio Figueiredo Guerra. *Aspectos Jurídicos Do Estudo de Impacto Ambiental (EIA).* [Legal Aspects of Environmental Impact Assessment (EIA)] MP Editora. São Paulo, 2008.

Benjamin, Antonio Herman. "Os Princípios do Estudo de Impacto Ambiental como limites da discricionariedade administrativo." [The Principles of the Environmental Impact Assessment as Limits of the Administrative Discretion]. *Review Forense* 317 (1992): 25–45. http://www.sidalc.net/cgi-bin/wxis.exe/?IsisScript=FARN.xis&method=post&formato=2&cantidad=1&expresion=mfn=000906.

Bermann, Celio "Governance of Infrastructure Investment in Brazil: Implications of Energy Related Mega Projects for Society, the Environment and Democracy." In *Infrastructure: For People or For Profit?* edited by Nancy Alexander, 45–57. Heinrich Boell Foundation North America, 2014.

Bishop, Cheryl Ann. *Access to Information as a Human Right.* El Paso: LFP Scholarly Pub. 2012.

Bom Dia Brasil, "Desmatamento Aumenta na Região da Usina de Belo Monte" [Deforestation Increases at the Belo Monte Dam Region], *G1*, October 27, 2015 8:53 a.m., http://g1.globo.com/bom-dia-brasil/noticia/2015/10/desmatamento-aumenta-na-regiao-da-usina-de-belo-monte-diz-instituto.html.

Bond, Alan, Thomas B. Fischer, and Josh Fothergill. "Progressing Quality Control in Environmental Impact Assessment Beyond Legislative Compliance: An Evaluation of the IEMA EIA Quality Mark Certification Scheme." *Environmental Impact Assessment Review* 63 (March 2017): 160–171. https://doi.org/10.1016/j.eiar.2016.12.001.

Brandt Meio Ambiente, *Estudo de Impacto Ambiental do Projeto Volta Grande* [Environmental Impact Statement of the Volta Grande Project] Item 5, 2010, (on file with author).

Brandt Meio Ambiente, *Relatório de Impacto Ambiental: Projeto Volta Grande* [Environmental Impact Assessment Non-Technical Summary: Volta Grande Project] (March 2012). https://www.sema.pa.gov.br/download/2BSML001-1-EA-RIM-0002_RIMA_REVISADO.pdf.

Branis, Martin. "A System of Certified Environmental Impact Assessment Experts in the Czech Republic." *Environmental Impact Assessment Review* 14, n. 2–3 (March–May 1994): 203–208.

Bratman, Eve. "Passive Revolution in the Green Economy: Activism and the Belo Monte Dam." *International Environmental Agreements: Politics Law & Economics* 15, n. 1 (2015): 61–77.

Bratman, Eve and Cristiane Bená Dias. "Development Blind Spots and Environmental Impact Assessment: Tensions between Policy, Law and Practice in Brazil's Xingu River Basin." *Environmental Impact Assessment Review* 70 (Feb. 2018): 1–10.

"Brazil," Environmental Democracy Index, last accessed September 15, 2019. http://environmentaldemocracyindex.org/country/bra.

"Brazil," OECD Better Life Index, last accessed September 15, 2019. http://www.oecdbetterlifeindex.org/countries/brazil/.

Bridgeman, Natalie L. and David Hunter. "Narrowing the Accountability Gap: Toward a New Foreign Investor Accountability Mechanism." *Georgetown International Environmental Law Review* 20, n. 2 (Winter 2008): 187–236. https://digitalcommons.wcl.american.edu/cgi/viewcontent.cgi?referer=https://www.google.com/&httpsredir=1&article=1066&context=facsch_lawrev.

Burdge, R.J., Fricke, P., and Finsterbusch, K. Sun. "Guidelines and Principles for Social Impact Assessment." *Environmental Impact Assessment Review* 15, n. 1 (1995): 11–43.

Caldeira, Vanessa Alvarenga. "Povos indígenas, sociedade moderna e projetos (ditos) de desenvolvimento nacional." *Tessituras: Revista de Antropologia e Arqueologia* 3, n. 1 (Jan–June 2015): 170–190. https://periodicos.ufpel.edu.br/ojs2/index.php/tessituras/article/view/5556/4129.

Cândido Rocha, Ednaldo et al., "Avaliação de Impactos Ambientais nos Países do Mercosul." [Environmental Impact Assessment in the Mercosur Countries], *Ambiente & Sociedade* 8, n. 2 (Jul–Dec. 2005): 147–160.

CARE Food and Nutrition Security, Pathways Program: The Talking Book, last accessed February 20, 2020. https://worldbankgroup.sharepoint.com/sites/gsg/CDD/Lists/Calendar/Attachments/209/CARE%20Talking%20Book%20Innovation%20Brief.pdf.

Carmo, Aline Borges. "A dimensão política do atual processo de agilização do licenciamento ambiental no Brasil." [The Political Dimension of the Current Process of Changing Environmental Licensing in Brazil to a Faster Procedure] *Revista Gestão & Políticas Públicas* 3, n. 2 (Dec. 2013): 284–306.

Carmo, Aline Borges and Alessandro Soares da Silva. "Licenciamento ambiental federal no Brasil: perspectiva histórica, poder e tomada de decisão em um campo em tensão." [Federal Environmental Permitting in Brazil: Historical Perspective, Power and Decision-making in a Field Experiencing Tension], *Revista Franco-brasileira de Geografia* 19, n. 19 (Nov. 2013): 1–18. https://doi.org/10.4000/confins.8555.

Castro Coura, Alexandre and Avance Rocha, Lívia. "Tutela Jurisdicional do Meio Ambiente, e Fenômeno da Colisão de Direitos Fundamentais na Jurisprudência do STF e do STJ" ["Jurisdictional Protection of the Environment, and Phenomenon of

Collision of Fundamental Rights in the Jurisprudence of the Supreme Court and the Superior Tribunal of Justice"], *Revista de Direito Ambiental* 19, n.76 (Oct. –Dec. 2014): 37–61.

Cavalcanti, Clovis. "Economic Growth and Environmental Protection in Brazil: An Unfavorable Trade-off." In *Environmental Governance and Decentralization*, edited by Albert Breton et al., 49–76, Cheltenham: Elgar Publishing, 2007.

"Cidades, Maranhão, Belágua" [Cities, Maranhão, Belágua], Instituto Brasileiro de Geografia e Estatistica (IBGE), last accessed May 25, 2015, http://cidades.ibge.gov.br/xtras/temas.php?lang=&codmun=210173&idtema=16&search=||s%EDntese-das-informa%E7%F5es.

"Cidades, Pará, Senador José Porfírio" [Cities, Pará, Senador José Porfírio], Instituto Brasileiro de Geografia e Estatistica (IBGE), last accessed October 14, 2019, http://cidades.ibge.gov.br/xtras/temas.php?lang=&codmun=150780&idtema=16&search=para|senador-jose-porfirio|sintese-das-informacoes.

Chaturvedi, Ipshita "A Critical Study of Free, Prior and Informed Consent in the Context of the Right to Development – Can Consent be Withheld?" *Journal of Indian Law & Society* 5 (Winter 2014): 37–60.

Code of Practice, Consultation in Ontario's Environmental Assessment Process, Ontario (website), last updated January 2014. https://www.ontario.ca/document/consultation-ontarios-environmental-assessment-process.

Coelho, Maria Célia Nunes, Luiz Jardim Wanderley, and Reinaldo Corrêa Costa. "Extrativismo do Ouro no Século XXI. Exemplos no Sudoeste da Amazônia Brasileira." [Gold Extrativism in the XXI Century. Examples at the Southeast of the Brazilian Amazon]. *Anuário do Instituto de Geociências* 39, n. 3 (2016): 5–14. http://www.anuario.igeo.ufrj.br/2016_3/2016_3_05_14.pdf.

Compliance Advisor Ombudsman, "About the CAO: Who We Are," last accessed March 22, 2016. http://www.cao-ombudsman.org/about/whoweare.

Compliance Advisor Ombudsman. *CAO Operational Guidelines,* 2013. http://www.cao-ombudsman.org/howwework/documents/CAOOperationalGuidelines2013_ENGLISH.pdf.

Conrad, Elizabeth, L. F. Cassar, M. Christie, and I. Fazey. "Hearing but Not Listening? A Participatory Assessment of Public Participation in Planning." *Environment and Planning, Government and Policy*, 29, n. 5 (Jan. 2011): 761–782. https://doi.org/10.1068%2Fc10137.

Convention for the Protection of the Natural Resources and Environment of the South Pacific Region (Noumea Convention), November 25, 1986, 26 I.L.M. 38 (1987).

Convention on Environmental Impact Assessment in a Transboundary Context, February 25, 1991, 309 U.N.T.S. 1989.

Constituição Federal [C.F.] [Constitution] (Braz.).

Corburn, Jason. "Bringing Local Knowledge into Environmental Decision Making: Improving Urban Planning for Communities at Risk." *Journal of Planning Education and Research*, 22, n. 4 (June 2003): 420–433.

Corwin, Miles "The Oil Spill Heard 'Round the Country." *L.A. Times*, January 28, 1989, http://articles.latimes.com/1989-01-28/news/mn-1162_1_oil-spill/2.

Council on Environmental Quality (CEQ) Regulations for Implementing the Procedural Provisions of the National Environmental Policy Act, 40 C.F.R. pts. 1500–1508 (2015) (2020).

Craik, Neil. *The International Law of Environmental Impact Assessment: Process, Substance and Integration*. Cambridge: Cambridge University Press, 2008.

Creighton, James L. *Involving Citizens in Community Decision Making: A Guidebook Program for Community Problem Solving*. Washington, DC: Program for Community Problem Solving, 1994. https://www.participatorymethods.org/resource/involving-citizens-community-decision-making-guidebook.

"Criteria for 'Better' Environmental Decision-making," CSERGE – ESRC Programme on Environment, Global Development Research Center, last accessed January 6, 2017. https://www.gdrc.org/decision/fs-criteria.html.

Cuppen, Miriam, Bertien Broekhans, and Bert Enserink. "Public Participation in EIA and Attitude Formation." *Impact Assessment and Project Appraisal* 30, n. 2 (April 2012): 63–74. https://doi.org/10.1080/14615517.2012.660348.

Daes, Erica-Irene A. Special Rapporteur, "Indigenous People and their Relationship to Land," UN. Doc. E/ECN. 4/Sub.2/2001/21 (June 2001): 49–50.

Davis, Rachel and Daniel M. Franks. *Costs of Company-Community Conflict in the Extractive Sector*. Cambridge, MA: Harvard Kennedy School, 2014. https://www.csrm.uq.edu.au/media/docs/603/Costs_of_Conflict_Davis-Franks.pdf.

Decreto No. 40, Octubre 30, 2012, Diario Oficial [D.O.] (Chile).

Decreto No. 88.351, de 1 de Junho de 1983, Diário Oficial da União [D.O.U.] de 03.06.1983 (Braz.).

Decreto No. 143, de 20 de Junho de 2002, Diário Oficial da União [D.O.U.] de 21.06.2002 (Braz.).

Decreto No. 6.040, de 7 de Fevereiro de 2007, Diário Oficial da União (D.O.U.) de 08.02.2007 (Braz.).

Decreto No. 99.274, de 6 de Junho de 1990, Diário Oficial da União [D.O.U.] de 07.06.1990 (Braz.).

Decreto No. 1320, del 13 de Julio de 1998, Diario Oficial No. 43.340, del 15.07.1998 (Colombia).

Deelstra, Y., S. G. Nooteboom, H. R. Kohlmann, J. Van den Berg and S. Innanen. "Using Knowledge for Decision-making Purposes in the Context of Large Projects in the Netherlands" *Environmental Impact Assessment Review* 23, n. 5 (Aug. 2003): 517–541. https://doi.org/10.1016/S0195-9255(03)00070-2.

"Delcidio diz que Dinheiro da Obra de Belo Monte Abasteceu Campanha" [Delcidio Says that Money from Belo Monte Implementation Assisted Presidential Campaign], G1, March 12, 2016, 11:03 p.m., http://g1.globo.com/jornal-nacional/noticia/2016/03/delcidio-diz-que-dinheiro-da-obra-de-belo-monte-abasteceu-campanhas.html.

Delong, Daniel, and Michael Fox. "Small Communities and Public Participation in the Cumulative Environmental Impact Assessment Process: The Case of Little Bouctouche River, New Brunswick." *Journal of New Brunswick Studies/Revue d'études sur le Nouveau-Brunswick* 6, n. 1 (June 2015): 13–31. https://journals.lib.unb.ca/index.php/JNBS/article/view/23058.

Dendena, Bianca, and Stefano Corsi. "The Environmental and Social Impact Assessment: A Further Step towards an Integrated Assessment Process." *Journal of cleaner production* 108, A (Dec. 2015): 965–977. https://doi.org/10.1016/j.jclepro.2015.07.110.

Department of Electricity Development and Ministry of Water Resources, *Manual for Conducting Public Hearings in the Environmental Impact Assessment Process for Hydropower Projects Kathmandu.* Nepal: Private Sector Hydropower Development Project, 2004. http://www.doed.gov.np/documents/Manual-for-Conducting-Public-Hearings.pdf.

Departamento de Infra-estrutura de Transporte (DNIT), Estudos de Viabilidade Técnica, Ecônomica e Ambiental [Technical, Economic and Environmental Feasibility Study, Department of Infrastructure and Transportation – DNIT], last accessed March 3, 2016. http://www.dnit.gov.br/download/planejamento-e-pesquisa/planejamento/estudos-de-viabilidade/lcs-097-2010-solic-public-texto-evtea-site-dnit.pdf.

Derani, Cristiane. *Direito Ambiental Econômico* [Economic Environmental Law]. São Paulo: M. Limonad, 3rd ed., 2008.

Directive 2014/52/EU of the European Parliament and of the Council of 16 April 2014 Amending Directive 2011/92/EU on the Assessment of the Effects of Certain Public and Private Projects on the Environment, 2014 O.J. (L 124) 1.

Doelle, Meinhard, and A. John Sinclair. "Time for a New Approach to Public Participation in EA: Promoting Cooperation and Consensus for Sustainability." *Environmental Impact Assessment Review* 26, n. 2 (March 2006): 185–205. https://doi.org/10.1016/j.eiar.2005.07.013.

Do Valor, "Obras de Pólo Petroquimico do Rio São Interrompidas por Anulação de Licenças," [Licenses Authorizing Implementation of the Petrochemical of Rio are Canceled], Folha de São Paulo, May 15, 2013, 9:24 a.m. http://www1.folha.uol.com.br/mercado/2013/05/1278907-obras-do-comperj-da-petrobras-sao-interrompidas-por-anulacao-de-licencas.shtml.

Duprat, Deborah. "A Convenção n. 169 da OIT e o Direito a Consulta Livre, Prévia e Informada" [The ILO Convention No. 169 and the Free, Prior and Informed Consent], In *Convenção No. 169 da OIT e os Estados Nacionais* [Convention ILO No. 169] Edited by Deborah Duprat, 53–78. Brasília, DF: Ministério Público da União, 2015. http://www.mpf.mp.br/atuacao-tematica/ccr6/documentos-e-publicacoes/artigos/docs_artigos/convencao-169-da-oit_web.pdf.

"Dutch Legislation," Netherlands Commission for Environmental Assessment, last accessed January 14, 2017, http://www.commi.

"Egypt," Netherlands Commission for Environmental Assessment, last updated August 28, 2019, http://www.eia.nl/en/countries/af/egypt/eia.

"EIA Q-Mark," Institute of Environmental Management and Assessment (IEMA), last accessed October 22, 2019, http://www.iema.net/eia-quality-mark.

"Environmental Impact Assessment-EIA," European Commission, last accessed September 17, 2015, http://ec.europa.eu/environment/eia/eia-legalcontext.htm.

"Environmental Impact Assessment," Environmental Protection Agency, last accessed October 22, 2019, http://www.epa.ie/monitoringassessment/assessment/eia/.

"Environmental Justice," United States Environmental Protection Agency, last accessed January 15, 2017, https://www.epa.gov/environmentaljustice.

Environmental Law Institute, Report of the International Roundtable on Environmental Impact Assessment and Public Participation in Environmental Decision-Making (1992) (on file with author).

Environmental Protection Agency, *Improved Science-based Environmental Stakeholder Processes.* Washington, DC: United States Environmental Protection Agency, 2001. https://yosemite.epa.gov/sab/sabproduct.nsf/cee3f362f1a1344e8525718e004ea078/$file/eecm01006_report_appna-e.pdf.

Environmental Resources Management, European Commission, *Guidance on EIA Scoping.* Luxembourg: Office for Official Publications of the European Communities, 2001. http://ec.europa.eu/environment/archives/eia/eia-guidelines/g-scoping-full-text.pdf.

"Fact Sheet: The Belo Monte Dam," Amazon Watch, last accessed November 23, 2019, https://amazonwatch.org/assets/files/2011-august-belo-monte-dam-fact-sheet.pdf.

Ferraz, Claudio *Electoral Politics and Bureaucratic Discretion: Evidence from Environmental Licensing and Local Elections in Brazil* (unpublished working paper, January 2007), https://pdfs.semanticscholar.org/4c65/b4c514acbe53aa7bd24250d2d5c8e6cab4d2.pdf.

Fiorillo, Celso Antonio Pacheco, Dione Mari Morita, and Paulo Ferreira. *Licenciamento Ambiental* [Environmental Licensing]. São Paulo: Saraiva, 2015.

Fluker, Shaun. "The Right to Public Participation in Resources and Environmental Decision-Making in Alberta." *Alberta Law Review* 52, n. 3 (2015): 567–579.

Fonseca, Willian and Omar Yazbek Bittar. "Critérios para Delimitação de Áreas de Influência em Estudo de Impacto Ambiental" [Criteria for Delimitation of Area of Influence in Environmental Impact Assessments] (Paper presented at First Brazilian Conference on Impact Assessment, October 15, 2012). https://www.scribd.com/document/353966779/CRITERIOS-PARA-DELIMITACAO-DE-AREAS-DE-INFLUENCIA-EM-ESTUDOS-DE-IMPACTO-AMBIENTAL.

Foratini, Gisela Environmental Licensing Director, Brazilian Institute of Environment and Renewable Natural Resources, Brazil, (Power Point Presentation, Licenciamento Ambiental Federal. 8º Encontro de Logística e Transportes. May 07, 2013) [Environmental Permitting Procedure. 8th Conference on Logistics and Transportation], https://sitefiespstorage.blob.core.windows.net/uploads/2013/05/8.30-GISELA-Forattini.pptx.

Foti, Joseph, Lalanath De Silva, Heather McGray, Linda Shaffer, Jonathan Talbot, and Jacob Werksman. *Voice and Choice: Opening the Door to Environmental Democracy.* Washington, DC: World Resources Institute, 2008. https://pdf.wri.org/voice_and_choice.pdf.

"Fundação Cultural Palmares" [Palmares Cultural Foundation], Governo Federal do Brasil, last accessed October 1, 2015. http://www.palmares.gov.br/?page_id=88&estado=RO.

Gabcikovo-Nagymaros Project (Hung. v. Slovk.), Judgment, 1997 I.C.J. 3 (February 5).

Ganson, Brian and Wennmann, Achim. *Business and Conflict in Fragile States: The Case for Pragmatic Solutions*. London: Routledge, 2016.

Gaventa, John and Gregory Barrett. "So What Difference Does It Make? Mapping the Outcomes of Citizen Engagement." *IDS Working Papers* 347 (Oct. 2010): 01–72. http://www.gsdrc.org/document-library/so-what-difference-does-it-make-mapping-the-outcomes-of-citizen-engagement/.

Giannasi, Carlos Alberto. "A Doutrina de Segurança Nacional e o Milagre Econômico." [The National Security Doctrine and the Economic Miracle]. PhD diss., São Paulo University, 2011. http://www.teses.usp.br/teses/disponiveis/8/8137/tde-30092011-170055/pt-br.php.

Gilpin, Alan. "Environmental Impact Assessment: Cutting Edge for the 21st Century." Cambridge, UK: University Press, 1995.

Glasson, John, and Nemesio Neves B. Salvador. "EIA in Brazil: A Procedures–Practice Gap. A Comparative Study with Reference to the European Union, and Especially the UK." *Environmental Impact Assessment Review* 20, n. 2 (April 2000): 191–225. https://doi.org/10.1016/S0195-9255(99)00043-8.

Glasson, John. "Principles and Purposes of Standards and Thresholds in the EIA Process." In *Standards and Thresholds for Impact Assessment*, edited by Michael Schmidt, John Glasson, Lars Emmelin and Hendrike Helbron, 3–17. Germany: Springer-Verlag Berlin Heidelberg, 2008.

Glasson, John, Riki Therivel and Andrew Chadwick. *Introduction to Environmental Impact Assessment*. London: Routledge, 3rd ed., 2012.

Gonçalves Egler, Paulo Cesar. "Improving the Environmental Impact Assessment Process in Brazil." PhD diss., School of Environmental Sciences University of East Anglia, 1998.

Goodland, Robert. "Free, Prior and Informed Consent and the World Bank Group." *Sustainable Development Law & Policy* 4, n. 2 (Summer 2004): 66–74.

Government of Canada, Impact Assessment Agency of Canada, Policy and Guidance, Impact Assessment Process Overview, Phase 1: Planning, last accessed January 18, 2020, https://www.canada.ca/en/impact-assessment-agency/services/policy-guidance/impact-assessment-process-overview/phase1.html.

Gotzmann, Nora. "Human Rights and Impact Assessment." Danish Institute for Human Rights, Matters of Concern Human Rights Research Paper Series, 2014. http://www.humanrights.dk/publications/human-rights-impact-assessment.

Grabner, Maria Luiza. "O Direito Humano ao Consentimento Livre, Prévio e Informado como Baluarte do Sistema Jurídico de Proteção dos Conhecimentos Tradicionais" [The Human Right to Free, Prior and Informed Consent as Bulwark of the Judicial System to Protect Traditional Knowledge]. *Boletim Científico ESMPU*. 45, n.14 (Jul.–Dec. 2015): 11–65. http://www.mpf.mp.br/atuacao-tematica/ccr6/documentos-e-publicacoes/artigos/docs_artigos/1-o-direito-humano-ao-consentimento-livre-previo-e-informado-1.pdf.

Greenpeace, *Barragens do Rio Tapajós: Uma Avaliação Crítica do Estudo e Relatório de Impacto Ambiental (EIA/RIMA) do Aproveitamento Hidrelétrico do Rio São Luiz do Tapajós* [Dams of the Tapajós River: A Critical Analysis of the Environmental Impact Assessment and its Non-technical Summary of the

Hydroelectric Use of the São Luiz do Tapajós River]. Brazil: Greenpeace, 2015, http://greenpeace.org.br/tapajos/docs/analise-eia-rima.pdf.

Greenspan, Emily. "Free, Prior, and Informed Consent in Africa: An Emerging Standard for Extractive Industry Projects," Oxfam America Research Backgrounder series, 2014. www.oxfamamerica.org/publications/fpic-in-africa.

Gregory, Robin and Timothy McDaniels. "Appendix B: Improving Environmental Decision Processes." In *Decision Making for the Environment: Social and Behavioral Science Research Priorities,* edited by Garry D. Brewer and Paul C. Stern. Washington, DC: National Academies Press, 2005.

Guerra, Sidney. "Competência Ambiental a Luz da Lei Complementar No. 140/2011" [Environmental Competence in Light of the Federal Law No. 140/2011], *Revista do Programa de Pós-Graduação em Direito da UFC,* 1 (2012): 125–140. http://www.periodicos.ufc.br/index.php/nomos/article/viewFile/377/359.

Günther, Markus, Gesa Geißler, and Johann Köppel. "Many Roads May Lead to Rome: Selected Features of Quality Control within Environmental Assessment Systems in the US, NL, CA, and UK." *Environmental Impact Assessment Review* 62 (Jan. 2017): 250–258. https://doi.org/10.1016/j.eiar.2016.08.002.

Hoffman, Rose Mirian. "Gargalos do Licenciamento Ambiental Federal no Brasil" [Shortcomings of the Federal Environmental Permitting in Brazil], Câmara dos Deputados (Jul. 2015). file:///C:/Users/cbenadias/Downloads/gargalos_licenciamento_hofmann.pdf.

Henisz, Witold J. et al., "Spinning Gold: The Financial Returns to Stakeholder Engagement." *Strategic Management Journal* 35, n. 12 (2014): 1727–1748.

Hunter, David, James Salzman and Durwood Zaelke. *International Environmental Law and Policy.* St. Paul: Foundation Press, 5th ed., 2015.

IBAMA, *Avaliação de Impacto Ambiental: Caminhos para o Fortalecimento do Licenciamento* Ambiental [Environmental Impact Assessment: Pathways to Enhance the Environmental Licensing] 2016, http://www.ibama.gov.br/phocadownload/noticias/noticias2016/resumo_executivo.pdf.

"Ibama Arquiva Licença Ambiental para Hidrelétrica São Luiz dos Tapajós" [Federal Environmental Agency – IBAMA – Nulls Environmental License Authorizing São Luiz dos Tapajós Dam], G1, August 5, 2010, 10:42 a.m., http://g1.globo.com/pa/santarem-regiao/noticia/2016/08/ibama-arquiva-licenca-ambiental-para-hidreletrica-sao-luiz-do-tapajos.html.

IBAMA, Technical Note sent to Norte Energia, November 2014 (on file with author).

IBAMA, Technical Note sent to Norte Energia, July 6, 2015 (on file with author).

Impact Assessment Act S.C. 2019, c. 28 (Canada). https://laws.justice.gc.ca/PDF/I-2.75.pdf.

Inage de Assis Oliveira, Antonio. *O Licenciamento Ambiental* [Environmental Licensing]. Brazil: Iglu, 1998.

INCRA Memorando [INCRA Memorandum] No. 0097/13 (on file with author).

Independent Environmental Monitoring Agency, last accessed December 8, 2019. https://monitoringagency.net/about/#the-agency.

Indigenous Communities of the Xingu River Basin, Pará, Brazil, Precautionary Measures, No. PM 382/2011, Inter-Am. Comm'n H.R. http://www.cidh.oas.org/medidas/2011.port.htm.

Instituto Socioambiental, *Análise do Licenciamento Ambiental do Projeto Volta Grande da Belo Sun Mineração* [Analysis of the Environmental Licensing of the Volta Grande Project owned by Belo Sun Mineração] 2013, https://www.socioambiental.org/sites/blog.socioambiental.org/files/blog/pdfs/2._manifestacao_belo_sun_isa.pdf.

Instituto Socioambiental and Rainforest Foundation Norway, *Consulta Livre, Prévia e Informada na Convenção 169 da OIT* [Free, Prior and Informed Consent in Light of the Convention ILO No. 169], last accessed February 22, 2020, https://site-antigo.socioambiental.org/inst/esp/consulta_previa/?q=convencao-169-da-oit-no-brasil#quem

"Intelligence Unit: Democracy Index 2018: Me too? Political Participation, Protest and Democracy 20." *The Economist*, last accessed September 15, 2019, http://www.eiu.com/Handlers/WhitepaperHandler.ashx?fi=Democracy_Index_2018.pdf&mode=wp&campaignid=Democracy2018.

Inter-American Development Bank, *Environment and Safeguards Compliance Policy*, January 19, 2006, http://idbdocs.iadb.org/wsdocs/getdocument.aspx?docnum=665902.

Inter-American Development Bank, *Operational Policy on Indigenous Peoples and Strategy for Indigenous Development*, February 22, 2006, http://idbdocs.iadb.org/wsdocs/getdocument.aspx?docnum=2032081.

International Association for Impact Assessment, *What is Impact Assessment?*, Oct. 2009, https://www.iaia.org/uploads/pdf/What_is_IA_web.pdf.

International Covenant on Economic, Social and Cultural Rights, Dec. 16, 1966, 993 U.N.T.S. 3.

International Finance Corporation, *Performance Standards*, 2012, http://www.ifc.org/wps/wcm/connect/1ee7038049a79139b845faa8c6a8312a/PS7_English_2012.pdf?MOD=AJPERES.

International Finance Corporation, *Policy on Environmental and Social Sustainability*, January 11, 2012, https://www.ifc.org/wps/wcm/connect/7141585d-c6fa-490b-a812-2ba87245115b/SP_English_2012.pdf?MOD=AJPERES&CVID=kiIrw0g.

Instrução Normativa do IBAMA No. 184, de 17 de Julho de 2008, Diário Oficial da União [D.O.U.] 18.07.2008 (Braz.).

Instrução Normativa do IBAMA No. 14, de 27 de Outubro de 2011, Diário Oficial da União [D.O.U.] de 28.10.2011 (Braz.)

Instrução Normativa do ICMBIO No. 14, de 27 de Outubro de 2011, Diário Oficial da União [D.O.U.] de 28.10.2011 (Braz.).

International Association for Impact Assessment, "The Linkages between Impact Assessment and the Sustainable Development Agenda, and Recommendations for Actions." Statements and Policy Briefing for the World Summit on Sustainable Development August 2002, http://www.iaia.org/pdf/IAIAMemberDocuments/News/WSSD/WSSDIAIA10pagerfinal0802.doc.

International Bank for Reconstruction and Development and International Development Association, *The World Bank Inspection Panel*, Resolution No. IBRD 93-10, September 22, 1993), https://www.inspectionpanel.org/sites/ip-ms8.extcc.com/files/documents/Resolution1993.pdf.

International Finance Corporation, *Stakeholders Engagement: A Good Practice Handbook for Companies Doing Business in Emerging Markets*. Washington, DC: International Finance Corporation, 2007, https://www.ifc.org/wps/wcm/connect/affbc005-2569-4e58-9962-280c483baa12/IFC_StakeholderEngagement.pdf?MOD=AJPERES&CVID=jkD13-p.

International Labor Organization. *Handbook for ILO Tripartite. Understanding the Indigenous and Tribal Peoples Convention, 1989 (ILO 169)*. Geneva: International Labor Office, 2013. https://www.ilo.org/wcmsp5/groups/public/---ed_norm/---normes/documents/publication/wcms_205225.pdf.

International Labour Organization (ILO), Indigenous and Tribal Peoples Convention, C169, 27 June 1989, C169.

Interviews with anonymous citizens and public officials involved with the environmental permitting of the Volta Grande Project, in Altamira, Senador José Porfirio and Belém, Pará State Brazil, July 2–9, 2015.

Jacobi, Pedro. "Ações da Belo Sun Fecham em 70c Após Mercado Reagir Mal ao Estudo de Viabilidade Econômica" [Shares of Belo Sun are at 70c After Market Had a Bad Reaction to the Economic Feasibility Study]. O Portal do Geólogo, May 7th, 2013 at 6:46 p.m., http://www.geologo.com.br/MAINLINK2013.ASP?VAIPARA=Acoes%C2%A0da%20Belo%20Sun%20fecham%20em%2070c%20apos%20mercado%20reagir%20mal%20ao%20estudo%20de%20viabilidade%20economica.

Jonathan, Fox, and Haight Libby. "Mexico's Transparency Reforms: Theory and Practice." In *Government Secrecy*, edited by Susan Maret, 353–379. United Kingdom: Research in Social Problems and Public Policy. Emerald Group Publishing Ltd, 2011.

J.F., ACP. No. 0001813-37.2014.4.01.3903, Subseção Judiciária de Altamira, 29.08.2014 (Braz.).

J.F., ACP No. 0001739-80.2014.4.01.3903, Subseção Judiciária de Altamira, 29.08.2014 (Braz.).

J.F., ACP No. 0002505-70.2013.4.01.3903, Subseção Judiciária de Altamira, 11.11.2013 (Braz.).

J.F. 5ª São Luis do Maranhão, Mandado de Segurança No. 2006.37.00.005222-7, São Luis do Maranhão 13.02.2007 (Braz.).

J.F. 1, Ação Civil Pública (ACP) No. 0006962-86.2014.4.013200, Amazonas, 23.05.2014 (Braz.).

Jones, Carys, Stephen Jay, Paul Slinn, "Environmental Assessment: Dominant or Dormant." In *Taking Stock of Environmental Assessment: Law, Policy and Practice*, edited by Jane Holder and Donald BcGillivray, London: Routledge-Cavendish, 3rd ed., 2007.

"Justiça Suspende Licença de Belo Monte Atendendo a Pedido do MPF" [Court Suspends the License of Belo Monte as Requested by Federal *Ministério Público*],

G1, August 31, 2016 10:27 p.m., http://g1.globo.com/pa/para/noticia/2016/08/justica-suspende-licenca-de-belo-monte-atendendo-pedido-do-mpf.html.

Klein, Peter Taylor. "Engaging the Brazilian State: The Belo Monte Dam and the Struggle for Political Voice." *The Journal of Peasant Studies* 42, n. 6 (March 2015): 1137–1156. https://doi.org/10.1080/03066150.2014.991719.

Klemens Laschefski, "Comunidades Tradicionais e Licenciamento Ambiental 'Entraves' para o Desenvolvimento" [Traditional Communities and Environmental Licensing: Are they Barriers to Development?], PowerPoint presentation, Environmental Licensing: Reality and Perspectives, Brasília, Nov. 5, 2015, http://www.mpf.mp.br/atuacao-tematica/ccr4/dados-da-atuacao/eventos/Seminarios/KlemensLaschefski.pdf.

Kolhoff, Arend J., Hens A. C. Runhaar, and Peter P. J. Driessen. "The Contribution of Capacities and Context to EIA System Performance and Effectiveness in Developing Countries: Towards a Better Understanding." *Impact Assessment and Project Appraisal* 27, n. 4 (Dec. 2009): 271–282.

Lawrence, David P. *Impact Assessment Practical Solutions to Recurrent Problems and Contemporary Challenges*. Hoboken, NJ: Wiley, 2nd ed., 2013.

Lei Complementar No. 140/11, de 8 de Dezembro de 2011, Diário Oficial da União [D.O.U.] de 9.12.2011 retificado em 12.12.2011 (Braz.).

Lei Estadual No. 5610 de 20 de Novembro de 1990, Diário Oficial do Estado do Pará [D.O.E.PA] de 20.11.90 (Braz.).

Lei No. 6.938, de 31 de Agosto de 1981, Diário Oficial da União [D.O.U.] de 02.09.1981 (Braz.).

Lei No. 7.347, de 24 de Julho de 1985, Diário Oficial da União [D.O.U.] de 25.07.1985 (Braz.).

Lei No. 1.356, de 3 de Outubro de 1988, Diário Oficial do Rio de Janeiro [D.O.E.R.J.] de 18.11.2008 (Braz.).

Lei No. 8.001, de 13 de Março de 1990, Diário Oficial da União [D.O.U.] de 14.03.1990 (Braz.).

Lei No. 9.605, de 12 de Fevereiro de 1998, Diário Oficial da União [D.O.U.] de 12.02.1998 (Braz.).

Lei No. 9.784, de 29 de Janeiro de 1989, Col. Leis Rep. Fed. Brasil, 1, 41, Fevereiro 1999 (Braz.).

Lei No. 10.650, de 16 de Abril de 2003, Diário Oficial da União [D.O.U.] de 17.11.2003 (Braz.).

Lei No. 11.284, de 2 de Março de 2006, Diário Oficial da União [D.O.U.] de 03.03.2006 (Braz.).

Leitão, Miriam "Delcidio Esclarece" [Delcidio, former Brazilian Senate, Clarifies], *O Globo* (blog), March 3, 2016, 8:30 a.m., http://blogs.oglobo.globo.com/miriam-leitao/post/delcidio-esclarece.html.

Leme Engenharia Ltda., *Estudo de Impacto Ambiental Belo Monte, Áreas de Influência* [Environmental Impact Statement of Belo Monte, Areas of Influence] May 2009. http://philip.inpa.gov.br/publ_livres/Dossie/BM/DocsOf/EIA-09/Vol%2005/AREAS%20DE%20INFLUENCIA/TEXTO/Cap%C3%ADtulo%206-%C3%81reas%20de%20Influ%C3%AAncia.pdf.

Leme Engenharia,"Relatório de Impacto Ambiental – RIMA da Hidrelétrica de Belo Monte" [Environmental Impact Statement Non-Technical Summary of Belo Monte Dam], May 2009, http://restrito.norteenergiasa.com.br/site/wp-content/uploads/2011/04/NE.Rima_.pdf.

Leme Machado, Paulo Affonso. *Direito Ambiental Brasileiro* [Brazilian Environmental Law] São Paulo: Malheiros 21st ed., 2013.

Ley No. 25.675, Nov. 27, 2002 (Argentina).

"Licenciamento Ambiental mais Rápido para Obras de Infraestrutura será Votado no Plenário" [Faster Environmental Licensing will be Subject to Vote at the Senate], Agência Senado, November 26, 2015, 10:26 a.m., http://www12.senado.leg.br/noticias/videos/2015/11/licenciamento-ambiental-mais-rapido-para-obras-de-infraestrutura-sera-votado-em-plenario.

MacGee, Brant "Participation with a Punch: Community Referenda on Dam Projects and the Right to Free, Prior, and Informed Consent to Development." *Water Alternatives* 3, n. 2 (2010): 162–184.

Mackay, Fergus "Indigenous Peoples and International Financial Institutions." In *International Financial Institutions and International Law*, edited by Daniel D. Bradlow and David B. Hunter, 287–320. Netherlands: Kluwer Law International, 2010.

Magalhães, Fernando Antonio Dorna. "Licenciamento Ambiental. Atuação do TCU." [Environmental Licensing. The Role of the Audit Court]. PowerPoint presentation, Environmental Licensing: Reality and Perspectives, Brasília, November 5th 2015, http://www.mpf.mp.br/atuacao-tematica/ccr4/dados-da-atuacao/eventos/Seminarios/Ppt0000053.pdf.

Marcelo, Fernando "Macaé: Realizada Audiência Pública das Termelétricas Jaci e Tupã," *Pre-Sal Info*, July 17, 2019, https://www.pre-sal.info/2019/07/17/macae-realizada-audiencia-publica-das-termeletricas-jaci-e-tupa/.

Maya Indigenous Communities of the Toledo District v. Belize, Case 12.053, Inter-Am. Comm'n. H.R., Report No. 40/04, Merits (October 12, 2004).

McAllister, Lesley K. "Environmental Advocacy Litigation in Brazil and the United States." *Journal of Comparative Law* 6, n. 2 (2011).

McComick, Sabrina. "The Governance of Hydro-Electric Dams in Brazil." *Journal of Latin American Studies* 39, n.2 (May 2007): 227–61.

Meirelles da Motta, Diana and Pêgo, Bolívar eds., *Licenciamento Ambiental para o Desenvolvimento Urbano: Avaliação de Instrumentos e Procedimentos* [Environmental Permitting Procedure for the Urban Development. Evaluation of Tools and Precedures]. Rio de Janeiro: IPEA, 2013.

Miazaki de Toledo, Thomaz. "Licenciamento Ambiental Federal" [Federal Environmental Licensing]. PowerPoint presentation, Environmental Licensing: Reality and Perspectives, Brasília, November 5th, 2015, http://www.mpf.mp.br/atuacao-tematica/ccr4/dados-da-atuacao/eventos/Seminarios/ThomazMiazaki.pdf.

Middle, Gary and Middle, Isaac "The Inefficiency of Environmental Impact Assessment: Reality or Myth?" *Impact Assessment & Project Appraisal* 28, n. 2 (Feb. 2012): 159–168.

Milano, Flavia and Sanhueza, Andrea *Public Consultations with Civil Society: Guidelines for Public and Private Executing Agencies*, March 2016, https://publications.iadb.org/handle/11319/7499.

Milaré, Édis. *Direito do Ambiente: A Gestão Ambiental em Foco* [Environmental Law: Environmental Management in Focus]. São Paulo: Revista dos Tribunais, 7th ed., 2011.

Miller, Janice. "Implementation of Principle 10 of the Rio Declaration of the United Nations Conference on Environment and Development." Paper, Caribbean Forum: Shaping a Sustainable Development Agenda to Address the Caribbean Reality in the Twenty First Century, March 5th, 2014.

"Mineradora Canadense Comprou Terras Irregulares na Amazônia" [Canadian Miner Bought Untitled Lands in the Amazon], *MAB Amazônia*, February 25, 2016 4:01 a.m., http://www.mabnacional.org.br/noticia/mineradora-canadense-comprou-terras-forma-irregular-no-xingu.

Ministério do Meio Ambiente, Instituto Brasileiro do Meio Ambiente e dos Recursos Naturais Renováveis, Banco Interamericano de Desenvolvimento and Programa das Nações Unidas para o Desenvolvimento, *Guia de Procedimentos do Licenciamento Ambiental Federal* [Federal Environmental Licensing Guide]. Brasília: Celaf, 2002. http://www.mma.gov.br/estruturas/sqa_pnla/_arquivos/Procedimentos.pdf.

Ministério do Meio Ambiente, *Áreas Prioritárias para a Conservação Uso Sustentável e Repartição de Benefícios da Biodiversidade Brasileira.* [Priority Areas to the Conservation and Sustainable Use and Sharing of Benefits of the Brazilian Biodiversity]. Brasília: Ministério do Meio Ambiente, 2007. http://www.mma.gov.br/estruturas/chm/_arquivos/biodiversidade31.pdf.

Ministério Público Federal, "Altamira: Audiência Pública sobre Belo Monte Marcada por Protestos" [Altamira: Protests Happened During Public Hearing to Present Belo Monte], International Rivers, September 15, 2009, 1:09 p.m., https://www.internationalrivers.org/node/1759.

Ministério Público Federal, *Processos Caso Belo Sun* [Lawsuits Against Belo Sun], last accessed November 24, https://docs.google.com/spreadsheets/d/1P4dXLzCpCPFNGlSarPFlkwDEygx3h4mgJmwVo_26sew/edit#gid=0.

Ministério Público Federal (MPF). "MPF Recomenda ao Governo do Pará que Suspenda todos os Licenciamentos na Volta Grande do Xingu" [Ministerio Público Recommends that the Government of Pará Suspends all the Environmental Permitting Procedures at the Xingu Volta Grande], last accessed November 23, 2019. http://www.mpf.mp.br/pa/sala-de-imprensa/noticias-pa/mpf-recomenda-ao-governo-do-para-que-suspenda-todos-os-licenciamentos-na-volta-grande-do-xingu.

Ministério Público de Minas Gerais, *Guia Técnico para Atuação do Ministério Público no licenciamento Ambiental de Atividade de Mineração* [Technical Guide for the Role of Public Prosecutors at the Environmental Permitting of Mining Activities]. Minas Gerais: Rev. do Ministério Público do Estado de Minas Gerais, 2012. www.mp.mg.gov.br/mpmgjuridico.

Ministério Público da União (4ª Câmara de Coordenação e Revisão - Meio Ambiente e Patrimônio Cultural), *Deficiências em Estudo de Impacto Ambiental* [Shortcomings of Environmental Impact Assessments] 2004.

Ministério do Meio Ambiente, *Manual de Normas e Procedimentos para Licenciamento Ambiental no Setor de Extração Mineral* [Handbook of Environmental Licensing Rules and Procedures of the Mineral Extractive Sector]. Brasília, DF: Ministério do Meio Ambiente, 2001. http://www.mma.gov.br/estruturas/sqa_pnla/_arquivos/MANUAL_mineracao.pdf.

Ministério do Planejamento. "As 10 Maiores Obras do PAC" [The 10 Biggest Projects of the Accelerated Growth Program] July, 2013, http://pac.gov.br/i/b8.

Environmental Impact Assessment Open Educational Resources, United Nations University, last accessed December 6, 2019, http://www.raymondsumouniversity.com/eia-local/page129.htm.

Morrison-Saunders, Angus, and John Bailey. "Practitioner Perspectives on the Role of Science in Environmental Impact Assessment." *Environmental Management* 31, n. 6 (June 2003): 683–695.

Morrison-Saunders, Angus, and Gerard Early. "What Is Necessary to Ensure Natural Justice in Environmental Impact Assessment Decision-Making?" *Impact Assessment and Project Appraisal* 26, n.1 (March 2008): 29–42.

Mox Plant (Ir. v. U.K.), ITLOS Case No. 10, Order of Dec. 3, 2001. https://www.itlos.org/fileadmin/itlos/documents/cases/case_no_10/Order.03.12.01.E.pdf.

"MPF Recomenda ao Conselho Nacional do Meio Ambiente Consulta à População sobre Licenciamento Ambiental" [Federal *Ministério Público* Recommends to the National Environmental Council Holding Public Consultation Regarding Environmental Licensing Proposal], Ministério Público Federal, March 7, 2016, 12:51 p.m. http://www.mpf.mp.br/pgr/noticias-pgr/mpf-recomenda-ao-conselho-nacional-do-meio-ambiente-consulta-a-populacao-sobre-licenciamento-ambiental.

National Environmental Policy Act of 1969, Pub. L. No. 91-190, 83 Stat. 852 (codified at 42 U.S.C. § 4321 (2012)).

Netherlands Commission for Environmental Assessment, *Keysheet: NCEA's Advice*, September 2011, http://api.commissiemer.nl/docs/cms/FS%2016%20Advisory%20procedure%20NCEA%20ENG%20Final.pdf.

Netherlands Commission for Environmental Assessment, *New Experiences on EIA in the Netherlands* 1999, http://api.commissiemer.nl/docs/mer/diversen/new_experiences_1998.pdf.

"Norway Halts Amazon Fund Donation in Dispute with Brazil," The Guardian, August 16, 2019, https://www.theguardian.com/world/2019/aug/16/norway-halts-amazon-fund-donation-dispute-brazil-deforestation-jair-bolsonaro.

Nuclear Tests (N.Z. v. Fr.), Judgment, 1974 I.C.J. 457 (Dec. 20, 1974).

OECD, "Recommendation of the Council on the Analysis of the Environmental Consequences of Significant Public and Private Projects," C (74) 216, November 14, 1974.

Okello, Nick, Lindsay Beevers, Wim Douven, and Jan Leentvaar. "The Doing and Un-Doing of Public Participation during Environmental Impact Assessments in Kenya." *Impact Assessment and Project Appraisal* 27, n. 3 (Sept. 2009): 217–226.

Orellana, Marcos A. "Democracia Ambiental y Desarrollo Sostenible: Hacia un Instrumento Regional sobre Derechos de Acesso" [Environmental Democracy and Sustainable Development: Towards a Regional Instrument on the Right to Access to Information] *FARN*: 43-61, 2014. https://accessinitiative.org/sites/default/files/orellana_iaf2014.pdf,

Ortolano, Leonard. "Controls on Project Proponents and Environmental Impact Assessment Effectiveness." *Environmental Professional* 15, n. 4 (Jan. 1993): 352–63.

Pagnoccheschi, B. and M. Bernado. "Política Ambiental no Brasil." [Environmental Politics in Brazil] In *Territorio, Ambiente e Políticas Públicas Espaciais*, edited by Marília Steinberger, 101–123. Brasília: LGE Publisher, 2006.

Pereira, Renee "Orçado em R$ 16 Bilhões, Custo da Usina de Belo Monte já Supera os R$ 30 Bilhões" [Belo Monte Dam Initial Budget of 16 Billion has already Surpassed 30 Billion Reais], *Estadão*, May 11, 2013 11:00 p.m., http://economia.estadao.com.br/noticias/geral,orcado-em-r-16-bilhoes-custo-da-usina-de-belo-monte-ja-supera-os-r-30-bilhoes,153398e.

Polido, Alexandra, and Tomás B. Ramos. "Towards Effective Scoping in Strategic Environmental Assessment." *Impact Assessment and Project Appraisal* 33, n. 3 (Jan. 2015): 171–183. http://dx.doi.org/10.1080/14615517.2014.993155.

Polito, Rodrigo. "Justiça Cassa Liminar que Suspendia Licença de Operação de Belo Monte" [Court Rules Injunction that Suspended Belo Monte Operation License], *Valor Econômico*, September 16th, 2016 at 9:18 a.m. http://www.valor.com.br/empresas/4711981/justica-cassa-liminar-que-suspendia-licenca-de-operacao-de-belo-monte.

Pölönen, Ismo. "Quality Control and the Substantive Influence of Environmental Impact Assessment in Finland." *Environmental Impact Assessment Review* 26, n. 5 (July 2006): 481–91.

Pölönen, Ismo, Pekka Hokkanen, and Kimmo Jalava. "The Effectiveness of the Finnish EIA System—What Works, What Doesn't, and What Could Be Improved?" *Environmental Impact Assessment Review* 31, n. 2 (March 2011): 120–128.

"Política de Responsibilidade Social e Ambiental" [National Development Bank, Social and Environmental Policy], Banco Nacional de Desenvolvimento Econômico e Social (BNDES), Oct. 2019, http://www.bndes.gov.br/wps/portal/site/home/quem-somos/responsabilidade-social-e-ambiental/o-que-nos-orienta/politicas/politica-responsabilidade-social-ambiental.

Portraria DNIT No. 1.705, de 14 de Novembro de 2007, Diário Oficial da União [D.O.U.] de 20.11.2007 (Braz.).

Portaria do ICMBIO No. 9, de 23 de Janeiro de 2007, Diário Oficial da União [D.O.U.], de 24.01.2007 (Braz.).

Portaria Interministerial No. 60, de 24 de Março de 2015, Diário Oficial da União [D.O.U.] de 25.03.2015 (Braz.).

Portaria Interministerial No. 419, de 26 de Outubro de 2011, Diário Oficial da União [D.O.U.] de 28.10.2011 (Braz.).

Portaria Interministerial No. 421, de 26 de Outubro de 2011, Diário Oficial da União [D.O.U.] de 28.10.2011 (Braz.).

Portaria Interministerial No. 422, de 22 de Outubro de 2011, Diário Oficial da União [D.O.U.] de 28.10.2011 (Braz.).

"Povos e Comunidades Tradicionais do Brasil" Nova Cartografia Social da Amazônia [New Social Cartography of the Amazon, Peoples and Traditional Communities of Brazi] (website) last accessed November 30, 2016. http://novacartografiasocial.com/fasciculos/povos-e-comunidades-tradicionais-do-brasil/?cp=2.

Press Release, Economic Commission for the Latin American and the Caribbean, "Fourteen Countries Sign New Generation Agreement at UN Headquarters on Access to Information, Public Participation and Justice in Environmental Matters." September 27, 2018. https://www.cepal.org/en/pressreleases/fourteen-countries-sign-new-generation-agreement-un-headquarters-access-information.

"Procedures," Netherlands Commission for Environmental Assessment, last accessed March 3, 2016, http://www.commissiemer.nl/english/our-services/types-of-advisory-report.

Procuradoria Geral da República, *A Área de Influência no EIA: Análise e Propostas* [The Area of Influence in Environmental Impact Assessments. Analysis and Proposals] 4CCR PGR, Nota Técnica 039/2007, October 17, 2007, http://www.mpf.mp.br/atuacao-tematica/ccr4/dados-da-atuacao/grupos-de-trabalho/gt-licenciamento/documentos-diversos/informa-tecnicas/NT%20039-07_Encaminha_trabalho_area_de_influencia.pdf/view.

"Project Complaint Mechanism," European Bank for Reconstruction and Development, last accessed November 5, 2019, https://www.ebrd.com/work-with-us/project-finance/project-complaint-mechanism.html.

Protocol on Environmental Protection to the Antarctic Treaty, arts. 6, 8, Oct. 4, 1991, 30 I.L.M. 1461.

"Public Participation Guide: Process Planning," United States Environmental Protection Agency, last accessed February 15, 2015, http://www2.epa.gov/international-cooperation/public-participation-guide-process-planning#step2.

Pulp Mills on the River Uruguay (Arg. v. Uru.), Judgment, 2010 I.C.J. 135 (Apr. 20).

Rajaram, T. and Ashutosh Das. "Screening for EIA in India: Enhancing Effectiveness Through Ecological Carrying Capacity Approach." *Journal of Environmental Management* 92, n. 1 (Jan. 2011): 140–148.

"Referendo" [Referendum], Tribunal Superior Eleitoral [Superior Electoral Tribunal], last accessed September 15, 2019, http://www.tse.jus.br/eleitor/glossario/termos/referendo.

Regional Agreement on Access to Information, Public Participation and Justice in Environmental Matters in Latin America and the Caribbean, March 4 2018, https://repositorio.cepal.org/bitstream/handle/11362/43583/1/S1800428_en.pdf.

"Regional Seas Program 1974," UN Environmental Project, last accessed November 16, 2019, https://www.unenvironment.org/explore-topics/oceans-seas/what-we-do/working-regional-seas/regional-seas-programmes/un-environment.

Relatório da Delegação do Brasil a Conferência das Nações Unidas sobre o Meio Ambiente – Volume II [Brazilian Delegation's Report to the United Nations Conference on the Human Environment – Volume II] (1972), https://cetesb.sp.gov.br/proclima/wp-content/uploads/sites/36/2013/12/estocolmo_72_Volume_II.pdf.

Resolução CEPRAM No. 2929, de 18 de Janeiro de 2002, Diário Oficial do Estado da Bahia [D.O.E.B.A.] de 18.01.2002 (Braz.).

Resolução CONAMA No. 001, de 23 de Janeiro de 1986, Diário Oficial da União [D.O.U.] de 17.02.1986 (Braz.).

Resolução CONAMA No. 006, de 24 de Janeiro de 1986, Diário Oficial da União [D.O.U.] de 17.02.1986 (Braz.).

Resolução CONAMA No. 009, de 03 de Dezembro de 1987, Diário Oficial da União [D.O.U.] de 05.07.1990 (Braz.).

Resolução CONAMA No. 237, de 19 de dezembro de 1997, Diário Oficial da União [D.O.U.], de 22.12.1997 (Braz.).

Resolução CONAMA No. 286, de 30 de Agosto de 2001, Diário Oficial da União [D.O.U.] de 17.12.2001 (Braz.).

Responsibilities and Obligations of States Sponsoring Persons and Entities with Respect to Activities in the Area, ITLOS Case No. 17, Advisory Opinion of Feb. 1, 2011, 10 ITLOS Rep. 10.

Retief, Francois, Angus Morrison-Saunders, Davide Geneletti, and Jenny Pope. "Exploring the Psychology of Trade-off Decision-making in Environmental Impact Assessment. *Impact Assessment and Project Appraisal* 31, n. 1 (March 2013), 13–23.

"Review and Update of the World Bank Safeguard Policies," World Bank, last accessed September 4, 2016. http://consultations.worldbank.org/consultation/review-and-update-world-bank-safeguard-policies.

Robertson v. Methow Valley Citizens Council, 490 U.S. 332 (1989).

Robinson, Raymond M. "Environmental Impact Assessment: The Growing Importance of Science in Government Decision Making." *Hydrobiologia* 188, n. 1 (Dec. 1989): 137–142.

Ross, William A. "The Independent Environmental Watchdog: A Canadian Experiment in EIA Follow-up." In *Assessing Impact: Handbook of EIA and SEA Follow-up*, edited by Angus Morrison-Saunders and Jos Arts, 178–192. London, UK: Routledge, 2004.

Runhaar, Hens, F. van Laerhoven, P. Driessen, and J. Arts. "Environmental Assessment in the Netherlands: Effectively Governing Environmental Protection? A Discourse Analysis." *Environmental Impact Assessment Review* 39 (Feb. 2013): 13–25.

Sanchez, Luis Enrique. *Avaliação de Impacto Ambiental: Conceitos e Métodos* [Environmental Impact Assessment Concepts and Methodologies], São Paulo: Oficina de Textos. E-BOOK, 2nd ed., 2013. https://ebooks.ofitexto.com.br/product/avaliao-de-impacto-ambiental-2-ed.

Sanchez, Luis Enrique. *Avaliação de impactos para licenciamento ambiental no Brasil: Modernização fundamentada no aprendizado.* [Environmental Assessment for Environmental Licensing in Brazil: Updates Based on Lessons Learned]. PowerPoint presentation, Environmental Licensing: Reality and Perspectives, Brasília, November 5, 2015. http://www.mpf.mp.br/atuacao-tematica/ccr4/dados-da-atuacao/eventos/Seminarios/Ppt0000055.pdf.

Sands, Philippe, Jacqueline Peel, Adriana Fabra and Ruth Mackenzie. *Principles of International Environmental Law*. Cambridge: Cambridge University Press, 3rd ed., 2012.

Santos Hoppe, Ariane, Maria Carolina Magedanz, Leni Dias Weigelt, Luciane Maria Schmidt Alves, Maristela Soares de Rezende, Aline Fernanda Fischborn and Suzane Beatriz Frantz Krug, "Participação Popular no Sistema Único de Saúde: Olhar de Usuários de Serviços de Saúde" [Public Participation in the Unified National Health Care System: Health Care Users' View], *Cinergis* 18, n. 1 (Dec. 2017): 1–9.

Saramaka People v. Suriname, Preliminary Objections, Merits, Reparations, and Costs, Judgment, Inter-Am. Ct. H.R. (Ser. C) No. 185 (November 28, 2007).

Sarkar, Rumu. "Exercising a Quasi-Judicial Review Through a World Bank Appellate Body." In *International Economic Law: The State and Future of the Discipline*, edited by Collin Picker Isabella Bunn and Douglas Arner. Oxford: Hart Publishing, 2008.

Sauer, Luke, Jaclynn Chiodini and Christine Duong. *Consent within Consultation. Incorporating New Business Practices in the Extraction Industry*, CIGI Graduate Fellows Series 3, 1 (July 2015): 1–8.

Sawakuchi, André Oliveira, Hartmann, G. A., Sawakuchi, H. O., Pupim, F. N., Bertassoli, D. J., Parra, M., Antinao, J. L.. "The Volta Grande Do Xingu: Reconstruction of Past Environments and Forecasting of Future Scenarios of a Unique Amazonian Fluvial Landscape." *Scientific Drilling* 20 (Dec. 2015): 21–32.

Secretaria Estadual de Meio Ambiente e Sustentabilidade, *Processo de Licenciamento Ambiental No. 2012/0000005028* [Environmental Permitting Procedure] (on file with author).

Servicio de Evaluación Ambiental, Qué es la Participación Ciudadana en el SEIA? [What is Public Participation in the Environmental Impact Assessment Service?], last accessed December 15, 2019, https://sea.gob.cl/participacion-ciudadana-y-consulta-indigena/que-es-la-participacion-ciudadana-en-el-seia.

Sharon, Kristin. *Environmental Justice: Creating Equality, Reclaiming Democracy.* Oxford: OUP, 2002.

Shen, Li-yin, Vivian W. Y. Tam, Leona Tam, and Ying-bo Ji. "Project Feasibility Study: The Key to Successful Implementation of Sustainable and Socially Responsible Construction Management Practice." *Journal of Cleaner Production* 18, n. 3 (Feb. 2010): 254–259.

Sinclair, A. John and Fitzpatrick, Patricia "Provisions for More Meaningful Public Participation Still Elusive in Proposed Canadian EIA Bill." *Impact Assessment & Project Appraisal* 20, n. 3 (Sept. 2002): 161–176.

Silva, Maria Amélia Rodrigues. "Royalties da mineração: instrumento de promoção do desenvolvimento sustentável de regiões mineradoras na Amazônia Oriental" [Mining Royalties, Instruments to Promote Development in the Mining Regions of the Eastern Amazon] *Novos Cadernos NAEA*. 1, n. 2 (1998). http://www.repositorio.ufpa.br:8080/jspui/bitstream/2011/3086/1/Artigo_RoyaltiesMineracaoInstrumento.pdf.

Silva Telles do Valle, Raul. "Sociedade Civil e Gestão Ambiental no Brasil: uma Análise da Implementação do Direito a Participação em nossa Legislação" [Civil Society and Enviromental Management in Brazil: An Analysis of the

Implementation of the Right to Participate in our Legislation]. Master's thesis, Universidade de São Paulo, 2002 (on file with author).

Silvestre de Carvalho, Lucas. "Análise de Stakeholders com Stratégia de Relacionamento Externo: Estudo de Caso de uma Empresa de Mineração em Minas Gerais" [Stakeholders Analysis with External Relationship Strategy: Case Study of a Mining Company in Minas Gerais] (Master's thesis, Universidade Federal de Lavras, 2013). http://repositorio.ufla.br/jspui/handle/1/1016.

"Sistema Informatizado de Licenciamento Ambiental" [Online System of Environmental Licensing], IBAMA, last accessed December 8, 2019). http://www.ibama.gov.br/sistemas/sislic/sobre-o-sislic.

Soltis, Katie. "Brazil Disregards Charges from the Inter-American Commission on Human Rights." *Council on Hemispheric Affairs*. June 9th, 2011. http://www.coha.org/brazil-disregards-charges-from-the-inter-american-commission-on-human-rights/.

Stampe, John W. "Lessons Learned from Environmental Impact Assessments: A Look at Two Widely Different Approaches–The USA and Thailand." *The Journal of Transdisciplinary Environmental Studies* 8, n. 1 (2009): 1–7. https://pdfs.semanticscholar.org/b8dc/b45ff29f0731ac659e9ebfe574d7d74445a1.pdf.

Susskind, Lawrence. "Multi-party Public Policy Mediation: A Separate Breed." *Dispute Resolution Management* 4, n. 1 (Fall 1997): 4–7.

"Terras Indígenas no Brasil" [Indigenous Lands in Brazil], Instituto Socioambiental, last accessed December 11, 2016, https://terrasindigenas.org.br/pt-br/brasil.

The Inspection Panel at the World Bank, *Operating Procedures* (April 2014), https://www.inspectionpanel.org/sites/ip-ms8.extcc.com/files/documents/2014%20Updated%20Operating%20Procedures.pdf.

The Mayagna (Sumo) Awas Tingni Community v. Nicaragua, Merits, Reparations and Costs, Judgment, Inter-Am. Ct. H.R. (ser. C) No. 66, ¶ 25 (Aug 31, 2001) citing The Mayagna (Sumo) Awas Tingni Community v. Nicaragua, Case 11.577, Inter-Am. Comm'n H.R., Report No. 27/98, ¶ 142 (Mar. 3, 1998).

The World Bank, *About the Inspection Panel*, last accessed March 29, 2016. http://ewebapps.worldbank.org/apps/ip/Pages/AboutUs.aspx.

The World Bank, *Environmental and Social Framework*, 99. Washington, DC: The World Bank, 2017, http://pubdocs.worldbank.org/en/837721522762050108/Environmental-and-Social-Framework.pdf.

The World Bank, *Operational Manual*, Operational Policy 4.01 (January 1999, revised April 2013), https://policies.worldbank.org/sites/ppf3/PPFDocuments/090224b0822f7384.pdf.

Tiwari, D. N., R. Loof, and G. N. Paudyal. "Environmental Economic Decision-making in Lowland Irrigated Agriculture Using Multi-Criteria Analysis Techniques." *Agricultural Systems* 60, n. 2 (May 1999): 99–112.

"Todos os Estados e o DF tem Protestos Contra o Governo Dilma" [Protests are Happening in All States and the Capital Against Dilma], G1, August 17, 2016, 6:47 a.m., http://g1.globo.com/politica/noticia/2015/08/manifestantes-protestam-contra-o-governo-dilma-em-cidades-do-brasil.html.

Tribunal de Contas da União, *Acórdão No. 2.212 Plenário, Relatório de Levantamento – Avaliação do IBAMA* [Report of Analysis of the Federal Environmental Agency – IBAMA] Sept. 23, 2009, https://pesquisa.apps.tcu.gov.br/#/resultado/acordao-completo/*/NUMACORDAO%253A2212%2520ANOACORDAO%253A2009/%2520.

Tribunal de Contas da União, *Acórdão No. 3413 Plenário, Licenciamento das UHEs de Jirau e Santo Antônio* [Environmental Permitting of the Hydroelectric Dams Jirau and Santo Antônio] Dec. 5, 2012. https://www.lexml.gov.br/urn/urn:lex:br:tribunal.contas.uniao;plenario:acordao:2012-12-05;3413.

Tribunal de Justiça do Estado de São Paulo, *Processos Distribuídos Justiça Estadual Brasileira - Dados de 2009 a 2013* [Lawsuits Filed at the State Courts – Data from 2009 to 2013]. Jan. 2015. (on file with author).

TRF-1, AI No. 0011749-24.2015.4.01.0000/Pa Pará, Relator: Des. Jirair Aram Meguerian, 12.03.2015 (Braz.).

TRF-1, A.I. No. 11749-24.2015.4.010000/Pa Pará, Relator: Des. Jirair Aram Megueriam, 24.04.2015 (Braz.).

Tunkin, Grigory "Is General International Law Customary Law Only?." *European Journal of International Law* 4, n. 4 (1993): 534–541.

Twelve Saramaka Clans (Lös) v. Suriname, Case 12.338, Inter-Am. Comm'n. H.R. Report No. 09/06, Admissibility and Merits (Mar. 2, 2006).

"UK Referendum," *BBC News*, last accessed September 15, 2019, https://www.bbc.com/news/politics/eu_referendum/results.

UN Conference on Environment and Development, Rio Declaration on Environment and Development, A/CONF.151/26/Rev.1 (Vol. I), Annex I (Aug. 12, 1992).

UNEP Governing Council, Goals and Principles of Environmental Impact Assessment: Decision 14/25, UNEP/GC/DEC/14/25 (January 16, 1987).

UNEP, "Training Manual," Environmental Impact Assessment Review Training Workshop, Kingstown, June 7–11, 2011, http://cep.unep.org/iwcam/documents/meeting-reports/environmental-impact-assessment-review-training-workshop.

UN Human Rights Office of the High Commissioner, *Guiding Principles on Business and Human Rights: Implementing the United Nations "Protect, Respect and Remedy" Framework*, A/HRC/17/31, 2011, http://www.ohchr.org/Documents/Publications/GuidingPrinciplesBusinessHR_EN.pdf.

United Nations Convention on the Law of the Sea, Dec. 10, 1982, 1833 U.N.T.S. 397.

United Nations Declaration on the Rights of Indigenous Peoples, A/RES/61/295 (Sept. 13, 2007).

United Nations Economic Commission for Europe, Convention on Access to Information, Public Participation in Decision-Making and Access to Justice in Environmental Matters, June 25, 1998, 2161 U.N.T.S. 447.

Urkiaga, A., L. De las Fuentes, B. Bis, E. Chiru, B. Bodo, F. Hernández, and T. Wintgens. "Methodologies for Feasibility Studies Related to Wastewater Reclamation and Reuse Projects Desalination." *Desalination* 187, n. 1–3 (Feb. 2006): 263–269.

U.S. Department of the Interior Bureau of Land Management, *Mount Hope Project Final Environmental Impact Statement Volume I of III*, October 2012, https://ep

lanning.blm.gov/epl-front-office/projects/nepa/76912/102398/125246/2012_MT_HOPE_FEIS_VOL_1.pdf.

"Usina a Fio D'agua e Menor Area Alagada" [Hydro Power Plant and Less Flooded Area], Usina Hidreletrica Belo Monte, Norte Energia, last accessed November 23, 2019, https://www.norteenergiasa.com.br/pt-br/uhe-belo-monte/arranjo.

Vanclay, Frank, Ana Maria Esteves, Ilse Aucamp, and Daniel M. Franks. *Social Impact Assessment: Guidance for Assessing and Managing the Social Impacts of Projects*. International Association for Impact Assessment, 2015.

Verdum, Ricardo. "Hydroelectric Power Invades Brazilian Indigenous Lands," *Americas Program* (blog), September 8th, 2015. http://www.cipamericas.org/archives/16168#.

Verocai, Iara "O Licenciamento Ambiental em Outros Paises" [Environmental Licensing in Other Countries]. Notes from Conference on Permitting Procedure, Environmental Protection and Development, July 29, 2004. http://www.mma.gov.br/estruturas/DAI/_arquivos/iaraverocai2.pdf.

Vidal de Souza, José Fernando and Erika Von Zuben. "O Licenciamento Ambiental e a Lei Complementar n.140/2011" [The Environmental Licensing and the Federal Law No. 140/2011], *Caderno de Direito* 12, n. 23 (2012): 11–44.

Vienna Convention on the Law of Treaties art. 27, May 23, 1969, 1155 U.N.T.S. 331.

Voto Escrito do Ministério Público sobre o Licenciamento Ambiental do Projeto Volta Grande [Written vote of *Ministério Público* on the Environmental Permitting Procedure of the Volta Grande Project] (on file with author).

Wasserman, Cheryl."Enforcement of Environmental Impact Assessment Requirements." Paper, Ninth International Conference on Environmental Compliance and Enforcement, June 2011. http://inece.org/conference/9/proceedings/57_Wasserman.pdf.

W.E. Westman, *Ecology, Impact Assessment and Environmental Planning*. New York: Wiley, 1985.

Weston, Joe. "EIA, Decision-making Theory and Screening and Scoping in UK Practice." *Journal of Environmental Planning and Management*, 43, n. 2 (2000): 185–203.

Wood, Christopher. *Environmental Impact Assessment: A Comparative Review*. London: Routledge, 2nd ed., 2003.

Wood, Christopher. "Environmental Impact Assessment in Developing Countries: An Overview." Paper presented at New Directions in Impact Assessment and Development: Methods and Practice, 2003. http://www.academia.edu/3420793/Environmental_impact_assessment_in_developing_countries_an_overview.

Wood, Christopher and Hartley, Nichola "Public Participation in Environmental Impact Assessment—Implementing the Aarhus Convention." *Environmental Impact Assessment Review* 25, n. 4 (May 2005): 319–340.

Wood, Graham, John Glasson, and Julia Becker. "EIA scoping in England and Wales: Practitioner Approaches, Perspectives and Constraints." *Environmental Impact Assessment Review* 26, n. 3 (April 2006): 221–241.

World Bank, *Environmental Licensing for Hydroelectric Projects in Brazil: A Contribution to the Debate*. March 2008. http://siteresources.worldbank.org/EXTWAT/Resources/4602122-1214578930250/Summary_Report.pdf.

World Economic Forum, *The Global Competitiveness Report 2015–2016: Brazil* (2016), http://reports.weforum.org/global-competitiveness-report-2015-2016/economies/#economy=BRA.

Yatsuda Moromizato Yoshida, Consuelo. "A Experiencia do Licenciamento Ambiental do Rodoanel Mario Covas e os Avancos no Trato das Questoes Federativas, Institucionais e Ambientais" [The Experience of the Environmental Licensing of the Mario Covas Beltway and the Progress Regarding the Federative, Institutional and Environmental Matters], in *Paisagem, Natureza e Direito* [Landscape, Nature and Law], ed. Antonio Herman Benjamin. 1 (2005): 231–250.

Zhouri, Andréa. "Justiça ambiental, diversidade cultural e accountability: desafios para a governança ambiental." [Environmental Justice, Cultural Diversity and Accountability] *Revista Brasileira de Ciências Sociais* 23, n. 68 (Oct. 2008): 97–107.

Zhouri, Andréa. "Mapping Environmental Inequalities in Brazil. Mining, Environmental Conflicts and Impasses of Mediation." Working paper no. 75, Desigualdades.net-Int. Res. Network on Interdependent Inequalities in Latin America, 2014. http://www.desigualdades.net/Resources/Working_Paper/WP-Zhouri-Online.pdf.

Index

Page references for figures are italicized

About the Author

Cristiane Bená Dias has more than a decade of experience with public participation in environmental and social impact assessments of projects in emerging economies. She has worked as an in-house attorney and environmental senior specialist at major private companies in Brazil. During that time, she conducted environmental and social due diligence and provided environmental legal assistance with respect to power plants, steel mills, ports, and mining operations. Currently, Dr. Bená Dias works at the Compliance Advisor/Ombudsman (CAO) for the International Finance Corporation (IFC) and Multilateral Investment Guarantee Agency (MIGA), the private financing arms of the World Bank Group. At the CAO, she investigates projects supported by IFC and MIGA to ensure compliance with their sustainability framework. IFC and MIGA finance projects in developing countries worldwide.

Dr. Bená Dias holds a Doctor of Juridical Science (SJD) degree focusing on comparative environmental law from American University Washington College of Law, United States, and a Master of Laws (LLM) in Public International Law from Queen Mary University of London, United Kingdom. Dr. Bená Dias is Brazilian and lives in Maryland, United States, with her family.

www.ingramcontent.com/pod-product-compliance
Ingram Content Group UK Ltd.
Pitfield, Milton Keynes, MK11 3LW, UK
UKHW020839290926
15470UKWH00004B/1450

9 781498 597746